D1483010

Winterland

Winterland

Create a Beautiful Garden for Every Season

CATHY REES

PHOTOGRAPHS BY LISA LOOKE

PRINCETON ARCHITECTURAL PRESS · NEW YORK

Contents

FOREWORD
by Rodney Eason
7

INTRODUCTION
Winter Garden Reflections
11

PART ONE
DESIGN
Revealing Garden Structure
14

DESIGN BASICS
PATHWAYS
FORM AND SCALE
TEXTURE
LAYERS
EVERGREENS
HEDGES
EDGES

PART TWO
CONTRAST
Emphasizing Polarities during the Season of Extremes
66

SUN / SHADOW
LIGHT / DARK
CONCEALED / REVEALED
UNIFORMITY / DIVERSITY
WILDNESS / CULTIVATION
MACRO / MICRO

PART THREE

EMBELLISH

Adorning the Garden for Year-Round Enjoyment

104

LIGHTING

ORNAMENT

STRUCTURES

STONE

PART FOUR

CARE

Creating a Big Impact through Small Acts

132

MAINTENANCE

DEER PROTECTION

PRUNING BASICS

SPECIAL PRUNING TECHNIQUES

PART FIVE

SHARE

Making the Most of Your Garden during the Quiet Season

160

BRINGING THE OUTDOORS IN

GETTING OUTSIDE IN WINTER

GARDENING FOR THE BIRDS

CREATING HABITAT

RESOURCES 188

ACKNOWLEDGMENTS 190

Foreword

Rodney Eason

Along the coast of Maine, in the northeastern arc of New England, the term *winter garden* is a euphemism. A euphemism for what exactly requires a bit of unpacking. Before I tackle the complexities and intricacies of the garden in winter, I must issue this important disclaimer: I grew up in the Piedmont of North Carolina, where I learned to garden in US Department of Agriculture (USDA) Plant Hardiness Zone 7b. There, a winter garden meant no walking on the frosty January grass until the sun came up and melted the frost away. After that, we could pretty much carry on as usual. The soil rarely froze, and when it did, it was just a thin patina on the outer mulch. Plants in Raleigh would flower or fruit year-round.

Winter is a good time to catch up on reading. One of my favorite gardening books in college was Christopher Brickell's *Encyclopedia of Gardening*. In this gorgeous book of photographs and how-to instructions, there was mention of "frost heave." What exactly was this? We never experienced frost heave in Zone 7b. The idea of a frost so deep and violent that it could make your plants heave out of the ground felt totally foreign to the garden's spaces of beauty and deep introspection. It reminded me of how I felt watching one of Wes Anderson's films—that I really did not understand it, but since the smart people were talking about it there must be something to it.

After two decades of pondering frost heaves, I now find myself trying to avoid them, as I live and garden in the Wes Anderson landscape of Mount Desert, Maine (USDA Zone 5b), near Acadia National Park. In the harsh coastal environment of Downeast Maine, I have learned firsthand that planting a perennial too late in the season will result in a rude rejection the following spring. The frost does not care if you bought the latest *Epimedium* from the horticultural emporium du jour or how enticing its catalog description might be; if it did not have enough time to set out good roots, once the snow melts in the spring, thirty dollars of desiccated stems and roots will lie prostrate on the ground, spat out of the earth with distaste.

The lead-up to winter is a sight to see. Fallen autumn leaves turn into a sepia blanket over the entire garden and lawn; in Mid-Coast Maine, the leaves fall in distinct waves for around a month: first the maples, then the alders, and finally the oaks. To keep up with all these leaves, I must rake the entire garden at least three times. If you are thinking, *Get a leaf blower*, my response is that raking is 100 percent eco-friendly, I love the exercise of raking, and I would much rather spend the money on plants than on a loud, fuel-guzzling leaf blower.

A recent winter came on brazenly and without warning, like a bully. One cold yet clear day in December, I went out to rake the last batch of leaves from our lawn. Just as I had assembled my piles, an icy rain started to fall, and I ran inside to the warmth of the woodstove. Overnight, the icy rain turned completely to ice and petrified my leaf piles. The ice then turned to snow, and slowly, I forgot about the leaves. When a rainstorm and a stretch of warm weather in late March melted the snow, there were my sad piles of leaves, waiting for me along with the frost-heaved perennials that had held so much hope at the end of the previous summer.

Some industrious type A gardeners like to keep themselves busy after a blanket of snow shuts down production for two to three months. They prune dead branches and remove unwanted trees while also shoveling snow. It is humorous to see someone shovel the pathways of their garden. And when I say someone, that someone is me. Shoveling snow is a good workout and an excuse to be outside, even when you cannot work in the soil or check on that overpriced *Epimedium* under its two-foot-thick white blanket. Shoveling snow for fun and inspiration does have a bit of cabin-fever-induced appeal, but I do also find that by clearing the paths, I can better see the form and function of the garden. When the sun is out, I get an idea of where it shines and where shade is cast by trees and shrubs. Are there certain stems and branches that need pruning or even removing? Winter is a good time to make notes about where to come back for cleanup in the spring, when I can actually feel my hands.

There is something about the dramatic stillness of the garden in winter that amplifies its intensity in summer. The season of cold, when we hear and feel the wind blow, makes me appreciate the necessity of the garden in my life. Gardeners want to be in the garden with our hands in the soil, to feel its warmth and to smell that earthy smell that only a rich garden soil can provide. We want to smell the spreading mint that we need to cull every year but just cannot bear to remove entirely. We look at the mass of *Chelone* stems and the grouping of *Deschampsia* and make notes to thin each. Or maybe, we tell ourselves, it would be best if the *Chelone* is moved over to that slightly shady spot that always stays kind of wet…

In our area of the world, which still experiences a profound and sometimes prolonged cold season, the winter garden is a time to take stock of what we have and what we have done, to find ways to keep warm while staying out in nature, and to dream of new plants, new pathways, and new features in our gardens. When I look at it from this perspective, I feel sad for all the gardeners back home in North Carolina, with their mild winter. They miss out on that dramatic change of seasons—and the deep feelings that the sleeping garden evokes.

Introduction

Winter Garden Reflections

Winter in the north: some folks flee to warmer places; others stick around, either by choice or necessity. As far as the garden is concerned, winter is the time after the leaves have fallen and before buds begin to break. Here in Maine, it is seven months long! Once this stunning fact penetrated my consciousness, I began to question why we put all our effort into gardening for the three magnificent but short months of summer. We often ignore our gardens between "putting them to bed" for the winter and beginning our spring chores. While I will be the first to admit that winter provides a much-needed break from gardening, that does not mean we need to put enjoyment of the garden on hold. Yes, it is dark out. Yes, it is cold. But winter can be glorious, too. Getting out into—or just admiring—the garden during this long season can enrich our lives and nurture our tenuous connection to nature.

Gardening for winter requires some different thinking. Unlike in summer, when you can rush out to the nursery, put your new plants into the ground, and begin appreciating them immediately, improving your garden for winter takes planning and the discipline to stick to that plan during the gardening season, when it may not seem so necessary. Observing and contemplating your garden while fully experiencing the period when the sun sets earlier and the temperatures rarely rise above freezing will inform your process.

As gardeners and designers, we are often encouraged to sketch out our scheme on paper, similar to how an architect makes drawings for a house. However, unlike the built environment, gardens are made up of living beings that do not remain static. They transform with the seasons and with time, in defiance of our rigid plans. How we cope with and manage changes in climate, weather, and, of course, the plants, sets true gardeners apart from those who garden out of habit or societal expectation.

While writing this book, I have visited many gardens. I found the most compelling to be the ones that have been developed over many seasons, according to the gardener's education and aesthetic. Far from following any formula, the owners tinker and experiment. The gardener and the garden evolve together over time, sometimes over a lifetime. This long-term evolution gives the gardener space to respond and grow along with the landscape. Relationships are formed, not just between gardener and garden but with individual plants, too. The life span of a perennial can be similar to that of a person; for a tree or shrub, it can be far longer. These relationships are what give the garden meaning, and the side effects are respect and beauty. Don't we all come to love what we know deeply?

People rely on rules, but every garden is a collaboration between the gardener and Mother Nature, and only one of us will be following them. All sites are unique; they each have their own history, resources, and challenges. Likewise, each gardener. There are no rules.

Instead of a prescription to make your garden more exciting, beautiful, or accessible in wintertime, I hope to give you some key design principles along with practices to help you implement them. I will encourage you to try new things and empower you to experiment and to adjust when something is not working—and, most of all, to learn as you go. *Gardening* is just another word for learning. Your teachers will be the plants and the rich community of organisms who call your garden home.

86

ONE

PART ONE

DESIGN

Revealing Garden Structure

Winter is a great time to reflect on your past year's gardening and to plan and design for the next season or the next several seasons. A garden is more a process than a thing, so it will never really be finished. You can go on pottering about and improving things for as long as you want, although it really does help to have some goals and a strategy to achieve them.

The design principles for creating a beautiful winter garden are much the same as those for any time of the year. The main difference is that in the winter, the structure of the garden is going to be much more visible. Without leaves and exuberant growth to distract or disguise, all you have to rely on is the skeleton. It is helpful to identify the backbone of that skeleton—structural elements such as paths, fences, outbuildings, and, of course, trees and shrubs—to determine if it is attractive and functional. The backbone influences how the garden will be accessed and provides a framework for what will be viewed and from where. It connects the house to the garden and anchors the garden in the greater landscape, establishing its place.

A garden that is composed mainly of lawn, or of herbaceous perennials or annuals, may not have much to show for itself in the winter, especially if everything is cut back in the fall. Adding elements that will form an attractive stand-alone backbone on which to hang summer foliage will enhance the beauty and functionality of your garden all year long. Winter is a perfect time to do this.

DESIGN BASICS

What kind of mood do you find appealing? Do you prefer formality or something more casual? There are some excellent reasons to go with a formal garden design: incorporating straight lines and simple geometric shapes confers a clean and orderly look, it is easy to lay out on the ground, and it may relate well to your home's architecture. Maintenance is straightforward, since a formal scheme tends to be more static and resistant to change. It is clear when something is out of place—a formal design can never be mistaken for nature—and it provides reassurance that the gardener has the upper hand.

On the downside, constructing a formal garden can be expensive if it relies on extensive hardscaping. Maintenance can be heavy; keeping the upper hand is not always easy. When a plant doesn't thrive, it will surely be noticed as one of many laid out in a regular pattern. Maintaining symmetry with live plants can be frustrating if site conditions are not uniform.

An entry garden where formal and informal garden elements complement the architecture of the house.

Naturalistically planted shrubs that reflect the care invested in their pruning and maintenance can feel formal in spite of their arrangement.

Informal gardens, on the other hand, make use of curved lines and asymmetry. A more naturalistic design may relate more readily to the landscape beyond the garden. The design can incorporate the varying soil and light conditions of your site. It can evolve and change as your plants thrive and grow. Developing a successful informal design, however, can be challenging: you will need to be in tune with your site and the surrounding landscape to make good choices.

A mixture of formal and informal garden spaces may work for your needs. Transitioning and incorporating elements from each in the other is essential to making a hybrid design harmonious. Consider formal layouts for areas closest to the house, where they can reflect the straight lines of the building, and informal ones beyond. Another great combination is formally laid out hardscaping softened with informal plantings. With plants alone, you can outline informally planted areas with geometrical features such as parterres surrounding blocks of meadow plantings for an orderly yet natural look.

An informal arrangement of conifers, shrubs, and grasses provides year-round color and texture.

Once you have decided on the overall style and mood of your garden, create a wish list of outdoor spaces and elements. We will discuss these elements and how to incorporate them into your garden throughout the book—but where should you start?

A garden overhaul may feel overwhelming, but you don't have to do it all at once. I recommend starting with the places you use and view most often in the winter. Strengthen and beautify the backbone, where you will appreciate the changes every day—the view from the kitchen or dining room or along your front walkway. Once you get a handle on those spaces, you can move outward and connect them. Don't forget to think about the sight lines that open up in the winter, exposing new views. Should these be highlighted or closed off? And while you may not be viewing the outside of

RIGHT Meticulous pruning results in a stylized version of nature.

BELOW RIGHT A completely artificial planting of fine-textured plants appears naturalistic, in harmony with small boulders and a random stepping-stone stairway.

BELOW Formal and informal elements mix in this walled garden laid out on two axes.

your home from the road that often, don't underestimate the impact of positive curb appeal, if only as you arrive home in the dark on winter evenings.

Capture photographs of your garden in winter, both the good and the not so good. Often the act of framing a view in a photograph helps you to see it better. Take photos from inside, through the windows from which you view the garden most often. Concentrate on the places you think need attention, and photograph from the places you are most likely to view them. Use these to develop a plan and refer to them often during the summer, when you are siting new plants and garden features, to make sure they will accomplish your winter goals. You can print out the photos or use editing software to draw or paste in trees, shrubs, fences, stone walls, et cetera, to help imagine how each view can be enhanced. A two-dimensional plan drawing just doesn't cut it for many people—and why struggle with measuring and plotting out a full-fledged plan if that method is not intuitive to you? You can improve your views one by one, using the frame of the photos.

By taking an incremental approach, you can learn as you go: plant in the summer, observe in the winter, and repeat. You might not want to apply this strategy to building a house, but it is quite acceptable and often most successful when building a garden. My only caution here is to keep the big picture in mind. Go out and experience your space in all seasons, so that what you are planning this year will not become a regret in the next. Do your research to make sure the plantings you are considering will survive in your climate, not outgrow their space, and fit well with each other as they mature.

Remember that gardening is a collaboration with Mother Nature. Just as every plant is part of nature, so are you. While gardening is often about exerting control to achieve an idealized version of nature, this does not have to be forceful, time consuming, or frustrating. By planning with nature in mind, you can eliminate a lot of maintenance and turn your ability to influence the natural world into an advantage. In most other art forms, we have a great deal of control over the outcome. Not so with gardening. Part of being a satisfied gardener is accepting the seasons, the soil, the insects, and the wildlife while still finding joy in change and your collaboration.

RIGHT An informal layout retains a bit of the wild.

BELOW A formal layout placing gates on a single axis feels predictable and comfortable.

PATHWAYS

Paths and walkways determine how, when, and which parts of the garden can be accessed and experienced. Enjoying the garden throughout the winter requires safe and easy passage, no matter what the weather has in store. Begin with the essentials, such as the walkways to your front door and to any other doors you normally use.

Does the line of the walkway relate to the style of the house and garden? Is it welcoming? Is the material durable, level, and easy to shovel? Is there a place to throw snow without damaging the plants? Is there something intriguing or beautiful to experience along this walkway during the winter?

A beautifully shaped walkway lined with plantings that transform with the seasons is a signal to visitors that they are being welcomed to the home of a gardener. The shape or line of the path should relate to the style of the garden while being as direct as possible. If no one uses the walkway from the street to your front door, consider changing its location; a more practical route from the driveway to the door may better serve you and your guests. The width and shape should be proportional to the house and entry. Narrow feels intimate, and wide feels grand; the width does not need to be the same from beginning to end. A walkway that becomes wider and enters a courtyard serves to welcome guests to an outdoor space before you greet them at the door; one that is wider at the point of entry can feel like a welcome mat.

Marking the beginning of the walkway with a structure such as a gate or trellis can give the visitor a feeling of entering a place apart. Trees or shrubs can do the same job. If you position two identical plants on either side of the walkway, make sure that light, moisture, and soil conditions are the same to encourage them to remain identical over time; pruning can also help to maintain symmetry. Add landscape lighting along your walkway to welcome guests during the dark hours of winter.

Choose plantings on either side of the walkway that have a presence year-round, but make sure they are tough if you will be shoveling snow onto them. Also, consider using sand along walkways during freezing weather as an alternative to salt. Sodium chloride, also known as road or rock salt, is the least expensive deicing alternative but the most problematic

ABOVE LEFT A rustic stairway with a natural wood railing feels comfortable leading to this cabin.

ABOVE RIGHT A casual stepping-stone path leads to the more formal entry from within an entrance courtyard.

LEFT Brick pavers in a straight line complement the symmetry of this front door.

LEFT A careful mix of paving materials adds an element of complexity.
RIGHT A brickwork pattern can be simple or intricate.

for your plants and soil. Foliage will turn brown when exposed to rock salt, and plants will appear burned after contact with salt or spray. Salt from previous applications will be shoveled onto foliage during snow removal and will be carried by melt water into the surrounding soil. Soil structure can be damaged, diminishing the water and nutrients available to your plants. While the safety of you and your family is most important, sand may be just as effective and can be swept up and reused. (Anything you use to counteract ice will make a mess when tracked indoors—sand is no worse than salt in that regard.)

Paths to other parts of the garden should be laid out with equal thought. Are there parts of the garden you visit more in the winter than in other seasons? Maybe the woodshed, or that big rock that warms up in the sun and provides shelter from the wind, or that lovely view that appears when the leaves fall? Daily trampling can be hard on a lawn if you need to get to your bird feeder or compost bin; these destinations warrant an official path. While the shortest distance between two points is a straight line, this might be most appropriate in a formal garden with an inherent geometry. Avoiding obstacles, obscuring a destination, winding past a

This unique path is made of bluestone pavers interspersed with small rounded rocks.
It is wider where it meets the driveway and narrows as it approaches the door.

TOP LEFT This path meanders through the garden before leading to the front door.

TOP RIGHT A break in the fence defines a pathway that leads beyond the line of sight.

BOTTOM LEFT A simple bridge provides access from gardened spaces to an informal woodland path.

cherished plant, or keeping a path on level ground are all good reasons to avoid straight lines. Creating a beautifully curved line is another. Since the path will contrast with its surroundings, whether they are beds, lawn, or woodland, that line will be a prominent visual feature. It will be viewed both as you walk along the path itself and as you look out your windows. Gentle, purposeful curves are usually best.

If your path traverses a low area, make sure there is adequate drainage or consider stepping stones or a bridge. Rain and meltwater move more slowly in the winter than in the summer. To give yourself time to troubleshoot, try out the route of your path for a season before committing to setting stones, pavers, bricks, or pea gravel. Once you determine the path, you can be creative in designing a pleasant journey, even if it is only to the compost bin!

A curving driveway obscures what lies ahead, enticing visitors.

The height and girth of a stately red oak (*Quercus rubra*) is accentuated by the open field that surrounds it.

FORM AND SCALE

Like all garden structure, plant forms are more prominent in the winter, when much of the garden's color and texture has fallen away with the leaves. While you can control the form of individual plants (or groups of plants, in the case of a hedge), each has its own natural form, influenced by its genetics, growth rate, branching habit, and external factors such as wind, nutrition, and water supply.

These factors come into play in the wild more than they do in a garden setting, where presumably the gardener is meeting plant needs and providing protection from browsing creatures. Since many of our traditional garden plants come from far away, it is hard to know how they would look in the wild. One thing is sure: a forsythia does not naturally grow in the shape of a ball or a cube!

Understanding the natural form of a plant can save a lot of effort in pruning and moving it from one place to another as it outgrows its original home. Some good references are available to describe the typical form and size of shrubs and trees (see Resources). It is particularly important to research this when thinking about plant arrangement: How will that tree look rising above that shrub? Will that ground cover play nicely with that shrub?

You can make an individual plant such as a small tree or beautifully pruned shrub a focal point by contrasting its form and color with its surroundings. You can also repeat forms throughout the garden to create rhythm. Columns, vases, and mounds are a few natural forms. *Twiggy*, *full*, *multistemmed*, *suckering*, *single-stemmed*, and *wispy* are some words that describe plant characteristics. Think about how to combine these types to highlight or contrast individual plants, how to meld smaller plants into larger units with greater mass and scale, and how to arrange diverse forms

Camperdown elm (*Ulmus glabra* 'Camperdownii') has a distinctive weeping habit.

RIGHT Small trees in the foreground echo those in the background, bringing the plantings into scale with the house and parking area.

BELOW LEFT The stone wall was built around this tree many years ago. Today, the big wall and big tree are in perfect scale with each other.

BELOW RIGHT Carefully tended plantings in scale with each other and the architecture make an inviting entrance.

An enclosed garden in scale with its surrounding stone wall. A large dogwood with distinctive arching shape complements the forms of the rhododendrons and boxwood beneath.

into a unified whole that stands alone without the benefit of intervening herbaceous perennials that will have died back before winter.

Every plant continues to grow over time but may slow down once it reaches a particular size. Because scale is such an important concept in the garden, planting and maintenance can be difficult. How can you plant using maximum size as a guideline when the plant is only a whip now? Planned senescence for some plants is one way. Spaces between trees and shrubs can be filled in with herbaceous plants until the woodies grow larger and shade out the intermingling plants. As your garden matures, you may need to take action to retain a pleasing scale of plants in relation to garden structures.

Pruning can go a long way in this regard. If you plant a fast-growing shrub in front of a slow-growing tree, anticipate problems by making sure

Bedrock Gardens in Lee, New Hampshire, is a celebration of art and horticulture. The repetition of trees in an allée on a long axis welcomes visitors in spite of the snowy weather.

you've chosen a shrub that will respond well to pruning until the tree gets the upper hand by shading the shrub and reducing its vigor. Keeping a shrub short by just trimming the top makes it look like it has had a bad haircut, which will be especially obvious in the winter. Instead, cut the tallest stems out completely so that new ones grow back, attaining their natural form. Pruning can also help maintain the scale of a tree that is threatening to dwarf your house. Regular pruning and shearing your conifers can postpone the day when they get too big for your yard.

If you already have a large-scale feature in your landscape such as a parking area or a meadow, balance it with large-scale plants (I'm talking trees here). A mature oak growing in a meadow is a beautiful sight! Use masses of a single plant or a tight mass of different plants to counterbalance a large feature like a garage or lawn.

Sculpture resonates with the natural forms of the surrounding forest.

ABOVE Winterberry holly (left) attains a vase shape without pruning; the paper birch (right) retains its lower side branches when it grows in the open.

RIGHT This large apple tree is an effective link between the garden and the wooded area beyond.

OVERLEAF Beatrix Farrand designed Garland Farm for her own year-round enjoyment. It has been restored and maintained by the Beatrix Farrand Society in Bar Harbor, Maine.

LEFT A beautifully maintained crabapple tree is the focus of this front yard and frames a distant view.
RIGHT This diminutive Japanese maple is bred to remain small, but may still need regular pruning to keep it in scale with nearby plantings.

If you inherited a garden of overgrown foundation plantings, think carefully about which plants you should eliminate and which you can control to complement the house and its surroundings. You might even try pruning some of them and waiting to see how they react before deciding which ones to keep.

External factors can have a big impact on form. Shade is an important one: a shrub that will become dense and full in the sun may take on a wispy, airy look when grown in the shade. Likewise, a plant normally found in a woodland understory will be taller, fuller, and more vigorous in the sun. Often, shade-loving plants can tolerate the sun if they are irrigated, but this too will contribute to their vigor. A plant that typically grows as a natural bonsai on the top of a mountain will be fuller and taller if you plant it in your fertile garden soil.

Learn as much as you can before planting trees and shrubs that you expect to appreciate for a long time and resist planting as though they will stay small forever. Give them some space and fill in with herbaceous plants until they mature into their role in the backbone of the garden. Design and prune with the knowledge that each plant has a natural form.

TEXTURE

Texture can be about tactility or about an irregular surface that is enhanced by the interplay of light and shadow. Tactility is generally reduced for us northerners in the winter, as we clothe ourselves in puffy mittens and insulated boots. Instead, we must perceive texture and contrast between surfaces with our eyes. We tend to focus on color when positioning plants in the garden, but texture should also be considered for winter, when colors are muted.

Varying the transparency of winter foliage can enhance complexity and interest while adding texture. A garden made up solely of solid evergreen forms will be boring unless a light snow or frost clings to needles, adding sparkle and a contrasting layer of white over the deep green. Many conifers have a dense profile that can be pruned into a more open shape, but there are limits to what pruning can accomplish. A combination of dense and airy makes for a more intricate and rewarding composition. Placing airy plants in front of a dense foliage backdrop creates a veil that increases mystery.

Combining grasses with evergreens produces a color as well as textural contrast. A tawny grass that catches the sun next to a deep-green conifer will highlight the best features of both. Placing perennials with attractive seed heads that can remain standing for the winter in front of a dense feature can also add depth and texture. Many seed heads turn a dark brown

OPPOSITE, LEFT Yellow birch (*Betula alleghaniensis*) has peeling bark that almost glows.

OPPOSITE, CENTER Many deciduous trees host lichen species that add color and texture to their bark.

OPPOSITE, RIGHT Korean stewartia (*Stewartia koreana*) has fantastic multicolored bark that peels off in plates.

RIGHT Perennials and grasses left standing add texture and depth to a winter garden while contrasting with evergreens in the background.

A variety of herbaceous and woody plants were selected to provide an abundance of color and texture.

or almost black as the season progresses, forming a rich tapestry of texture and color when combined with airy, straw-colored grasses. Or try contrasting deciduous shrubs, which become more transparent as they drop their leaves, with other plants or features. An airy shrub with russet-colored bark (such as spirea) will add depth and complexity in front of a stone wall.

Herbaceous perennials that disappear completely or turn to mush at the first frost will leave the area bare in the winter. That is not to say that every space needs to be filled, just that those remaining spaces between plants should be used in a conscious way. One winter, I left my oriental lilies standing instead of cutting back as usual. They looked great all winter—tall, sturdy, bleached stems, branched at the top. Some other

TOP Meadow plantings of perennials and grasses contrast with the path mown into a spiral.

CENTER The coarseness of the dwarf spruce contrasts with the fine texture of surrounding perennials.

BOTTOM LEFT Texturally similar plants are distinguished by their contrasting forms.

BOTTOM RIGHT The smooth texture of rounded stones laid on edge contrasts with the surrounding pea gravel.

plants that look great through the winter are grasses such as silver grass (*Miscanthus sinensis*), switchgrass (*Panicum virgatum* cultivars), and hair grass (*Deschampsia cespitosa*), or perennials such as Russian sage (*Perovskia atriplicifolia*), blue vervain (*Verbena hastata*), or purple coneflower (*Echinacea* spp.). Shrubs that retain their seed heads, such as steeplebush (*Spiraea tomentosa*), summersweet (*Clethra alnifolia*), or rhodora (*Rhododendron canadense*) or those that have horizontal branching, such as cotoneasters or viburnums, may suddenly take center stage after a light snow or frost. And don't forget about mosses such as haircap that evoke plush upholstery swathing the ground.

Think about using texture in the winter as you would use color in the summer: as a focal point, a contrasting element, a mass, or a repetitive element that creates rhythm as the eye takes in the quiet winter garden.

Once you have chosen, arranged, and positioned your plants to make the most of your winter garden, maintenance is the next step. Mowing and pruning can enhance texture by creating contrast. Try mowing only some sections of lawn or meadow, so that remaining foliage will be left at various heights, or pruning some plants into an open form and leaving others dense or mounding. The juxtaposition of textures will be heightened by the low angle of the sun or by a light snowfall or frost.

Variations on the same theme each produce a different textural effect.

Layers of plants that differ in color, texture, and form make this lovely scene complex and engaging.

Texture can also be found in the inanimate surfaces of the garden. Think about how you could use raked gravel, stones laid on edge, or a rustic or wattle fence to enhance visual texture. Or how about that blank wall on the garage? A trellis, with or without a climbing plant, can add shadows that shift throughout the day.

LAYERS

Made up of an indistinguishable number of species on a single horizontal plane, the ubiquitous lawn doesn't look too exciting in the winter, when it is brown or featureless, covered by snow. Like a blank canvas, a clean unbroken surface has a certain appeal, but how long can that hold your interest? On top of that, the minute someone tracks over it, that appeal vanishes.

I am lucky to live in the woods, and I have chosen to garden around the house rather than lawn around it. The surface of the ground is uneven, the

The surface of the shrub planting rises and falls, creating the illusion of topography.

result of thousands of years of trees tipping over, pulling up loads of soil with their roots, and then dropping them next to the hollows from which they were excavated. This uneven ground provides microhabitats for plants, mosses, and other creatures. It also provides topographical relief, the kind that can't be obtained with a bulldozer.

If your yard is flat and contains large expanses of lawn, you can make up for the lack of topographic relief by adding layers of plants. Trees, shrubs, and herbaceous plants each occupy a certain height range and, in doing so, form individual layers. Creating multiple layers is mainly a matter of choosing from the multitude of plant forms available: there are overstory trees and understory trees, and shrubs that have tall, arching forms or short, clonal, or sprawling forms. Among herbaceous plants, there are upright, nearly woody plants and there are ground covers. Don't forget the mosses. You get the idea: layers upon layers, building up a complexity that disguises the lack of topographic relief on the ground. Layered plantings also look fabulous in the summer!

To get a layered look, you don't need to immediately install three plants where, in the past, you thought one would do. You can build up the layers over time. When creating planting plans on paper, you can draw each layer on tracing paper to see how they overlay one another. Or you can use photographs that show the various views in elevation. Having all the layers visible at once in a single view can help you visualize the plants and how they relate to each other. Begin by planting the key shrubs and trees, then add in herbaceous ground covers over time. Another alternative is to plant the trees, some shrubs, and some herbaceous plants, then divide and fill in or expand each area after a couple of years. An incremental approach can help you get the job done without breaking the bank or crushing your motivation.

Undulating land and exposed boulders create growing conditions for a diverse selection of plants.

Massive white pines make up the canopy layer that overtops spruce and fir in the background, a stone wall in the middle ground, and boxwood in the foreground. Complex layering such as this can take a lifetime to achieve, but without the pines could be attained much sooner.

Layers of plants and stonework alternate to make the most of this roadside planting.

If you will need to remove existing vegetation such as a lawn, consider implementing your plan just a small section at a time or one layer at a time. I have found that the easiest, but not necessarily the quickest, way to remove a lawn is to load piles of leaves onto areas where you want to kill the grass in the fall. Leave them there until May or June, when you are ready to plant. Remove the leaves or rake them aside and dig your holes directly into the weakened lawn or a small mound of compost. You can either remove the sod to the compost pile or break it up and use it when backfilling the holes around your new trees and shrubs. Keep the leaves in place between the new plantings so they continue to smother the former lawn and keep applying leaves as a thick layer of mulch for the first year. This will also reduce your watering needs. The next year, when the leaves have decomposed a bit more, you can just move them aside and add plants and ground covers directly into the soil beneath. If you are planting smaller herbaceous plants or veggies that will easily be buried by the leaf mulch, you will have to flip the sod over and break it up with a spade, add compost or weed free topsoil and plant directly in that. Continue to reuse the leaves as much as possible to mulch your new planting and discourage remaining roots from resprouting.

Layering also works along the horizontal plane. Rather than thinking of your garden view as a two-dimensional composition that you might hang on the wall, you can use perspective to enhance depth and perception. Establishing layers in the foreground to frame a distant view, creating horizontal bands of structure to evoke a restful view, or adding vertical elements to accentuate the distance to a far-off view are all techniques you can play with to keep the scene engaging and changing over time. By view, I

Layers of plants play with scale—both dwarfed by the fence and overshadowing the fence—as this garden steps down a slope.

The vertical trunks of paper birch trees intersect terraces retained by stone walls. Cedars provide the backdrop to enhance this snowy scene at the Camden Public Library Amphitheatre designed by Fletcher Steele.

don't necessarily mean an expansive water or mountain vista; the view can be a short distance away and only a few feet wide. All you need is a focal point like the trunk of a large tree, a special boulder, or a sculpture that you want to accentuate.

Adding layers, either one above the other, or one in front of the other, will build up a richness that will be especially appreciated in the winter when everything looks sparse. Again, every layer doesn't need to be incorporated from day one. Developing layers takes time and thought. Make an ongoing plan to accomplish your objectives one layer or area at a time each winter. Observe the results as plants grow throughout the seasons and adjust your plantings as the scene develops.

A bed of mixed conifers is carefully pruned to maintain a tapestry of color and texture.

The foliage of this blue spruce (*Picea pungens*) is a wonderful accompaniment to a white picket fence.

EVERGREENS

As you contemplate plants for winter enjoyment, the first that come to mind may be evergreens. Conifers that do not lose all their needles in one season, like our native larch, boast a variety of deep-green, blue-green, or golden tones that look good all winter.

Think about placing your conifers in groups throughout the garden, as they may look lonesome dotted singly around when intervening herbaceous foliage is absent. Combining them with deciduous shrubs also works well. Conifers function nicely as a dark background for shrubs or small trees with light-colored bark such as shad bush (*Amelanchier* spp.), seven-son flower (*Heptacodium miconioides*), magnolia, viburnum, or deciduous holly. However, one single specimen can also make an impactful focal point.

Ornamental conifers come in a huge range of sizes and colors, thanks to a flurry of creative propagation in the past couple of decades. No longer limited to deep green and blue-green, many cultivars with silver and golden foliage are now available. Be aware that many take on a bronzy hue in the winter that you may either welcome or regret, so be sure to check this before buying. Cultivars have been selected to attain a variety of columnar, pyramidal, upright spreading, or sprawling forms. The tighter forms look best in a more formal arrangement but can make a nice counterpoint when used among other plants with loose or airy structures. Fantastical weeping and drooping conifers are also available. As with so many unusual plants, it can be easy to overdo, so use sparingly for impact or as a focal point.

Dwarf and miniature conifers can be great accents in a small garden or incorporated into a border mingled with deciduous shrubs and perennials. A miniature conifer will typically grow less than 1 inch (2.5 centimeters) per year, and a dwarf will put on from 1 to 6 inches (2.5 to 15.3 centimeters) per year. An intermediate conifer is expected to grow 6 to 12 inches (15.3 to 30.5 centimeters) per year, compared to a full-size plant, which will grow more than 12 inches (30.5 centimeters) per year. It is possible to delay the arrival of some full-size species in the overstory by regular shearing or pruning, but this can destroy some of their natural softness and form. All conifers continue to grow throughout their life, so even a dwarf can reach 15 feet (4.6 meters) after thirty years without pruning! Plan or maintain with this in mind. Make sure to plant conifers far enough from

LEFT Many evergreens such as these small-leaved rhododendron cultivars turn a shade of mahogany in the winter. CENTER A tapestry of fine-textured junipers. RIGHT False cypress (*Chamaecyparis obtusa* cultivars) retain their golden color throughout the winter.

Full-size conifers integrate a *torii* into its surroundings.

the house to accommodate ten or fifteen years' worth of growth. Planting too close is such a common problem that you can probably spot multiple examples by walking two blocks in any direction. By abandoning the conifer foundation-planting idea altogether and creating a conifer island, you can avoid this problem.

All conifers can be pruned to maintain their size and improve their shape, within reason. If you are new to pruning conifers, it would be wise to consult one of the many excellent books on pruning (see Resources for some suggestions). In general, it is best to prune most conifers when they are not actively growing. Always prune back to a bud or green shoot. While cedars and yews may resprout from below the green needles, older specimens are not likely to respond. With pines, however, the best time to prune is when the shoots at the tips are elongating in the early summer. Pinching off up to two-thirds the length of these shoots is called candling because of the appearance of the elongating bud before the needles begin to expand.

ABOVE An artful mix of evergreens and deciduous shrubs soften the hard edges of a stone.

TOP RIGHT Massive rhododendrons mix with a mugo pine and deciduous shrubs.

RIGHT A plant collector's mix of evergreens, including varieties of false cypress (*Chamaecyparis* spp.), mugo pines, and Siberian carpet cypress (*Microbiota decussata*).

Trying to keep large conifers in scale within a small landscape will require annual pruning; depending on the plant, it may be a losing battle. In my yard, where native red spruce, white pine, and balsam firs are sprouting up all the time, I think of them as a rotating crop, taking out the ones that are getting too big and letting the smaller seedlings grow up to take their place. I value their color and texture but don't need any more canopy trees near the house.

Many ornamental conifers have a very dense branching pattern that may be a result of how they were bred or how they were pruned at the nursery, before they came to you. Even if you intend to use conifers to form a screen, thinning out some of the branches can increase air flow, keep the plant healthy, and prevent it from turning into a blob. Wear long sleeves when pruning conifers to protect yourself from scratchy needles and avoid contact with junipers and microbiotas in particular, which can cause a rash. It is good practice to prune conifers on a dry day, so as not to encourage the spread of fungi that may be living on your plants. Disinfecting pruners with a diluted bleach solution or alcohol wipe between plants also helps prevent spreading fungi from one plant to another.

Dwarf and miniature conifers can be grown in containers. Make sure the container is weatherproof, has good drainage, and can be moved once full. The smaller, slower-growing conifers can live happily for many years in a container if kept moist but not overwatered. Make sure they are hardy a zone or two below yours, since the pot will not be providing the same insulation as the ground in the winter.

And don't forget about the broad-leaved evergreens such as rhododendrons of all shapes and sizes, evergreen hollies such as *Ilex × meserveae* and *I. glabra*, *Pieris*, mountain laurel (*Kalmia latifolia*), and *Leucothoe*. Low-growing bearberry (*Arctostaphylos uva-ursi*) and heather (*Calluna vulgaris* cultivars) are also great choices. Be aware that the leaves of many rhododendrons and laurels will droop when temperatures plummet. This is an adaptation to reduce water loss from their leaves when water cannot be replenished from the frozen ground. And, as with conifers, do your research to determine what the winter foliage color will be. Combining conifers and broad-leaved evergreens can make for a stunning winter scene.

A very old cedar hedge pruned above deer height.

A hedge of American hornbeam (*Ostrya virginiana*) retains its leaves, which flutter and add sound and movement to the winter garden.

HEDGES

Living elements of your garden structure, hedges and edges can be designed to lie anywhere on the continuum from formal to informal. A formal hedge consists of a single species in an uninterrupted line, regularly clipped to maintain a specified form that remains a static element in the landscape. Whether evergreen or deciduous, a formal hedge is basically a living wall or fence that will need pruning or shearing at least once per season, depending on your choice of plant. Geometric or wavy shapes may need monthly maintenance to look their best. Formal hedges are often used to separate one space from another within a garden, to enclose a space, or to screen an unwanted view.

Low hedges are useful along the perimeter of herb gardens or parterres. Beds are typically outlined with a single shrub species that is clipped to form a uniform edge in a geometric shape. Dense mounding shrubs like boxwood or lavender can be either shaped or left unclipped for a more

ABOVE LEFT Two pruned and shaped hemlocks signal an entry through a border of untrimmed trees.

ABOVE RIGHT A geometrically pruned cedar hedge makes a great foil for the mature tree in the background.

RIGHT The irregular line of this mixed shrub border creates an effective screen, lending privacy and obscuring the straight line of the street beyond.

Low hedges accentuate the outline of a zigzag walkway.

informal look. With a uniform outline, a sense of order will preside, whether the interior of the bed follows a formal plan or is just pure mayhem.

At the other end of the spectrum is an informal hedge, otherwise known as a hedgerow or shrub border. Usually, made up of mixed species and left unsheared (but not necessarily unpruned); they vary in height and continuity. They will serve much the same purposes as a formal hedge, but they can be used more creatively—as an antidote to a wide-open view or as a way to evoke a sense of mystery. While hedgerows traditionally occupy straight lines between fields, irregular lines usually work best if you are mixing species with a diversity of color, height, and texture. This type of shrub border can also effectively obscure an unwanted straight line such as

A mossy area and short retaining wall signal the boundary between a garden and adjacent woods.

a property boundary. Typically, a shrub border will be deeper and will take up more area than a more formal clipped hedge.

Variations along the spectrum from formal to informal and plant selections are limited only by the gardener's imagination. Cedar, hemlock, or yew are commonly used, but many more choices could be appropriate for a single-species formal hedge (such as lilac, winterberry, fothergilla, or hydrangea). Avoid privet, honeysuckle, burning bush, and any other plants that are considered invasive in your area. Also, make sure your selection can withstand regular shearing with grace; I do not recommend deciduous shrubs with large or compound leaves because these are difficult to shear into a compact shape. Choices are unlimited for a shrub border, since the natural form of each plant can easily be maintained by annual pruning. Combine a variety of shrubs to become a vertical tapestry of twig color and branching pattern in winter and a floriferous display in summer.

EDGES

Plantings can be used to integrate the garden into the larger landscape by creating an informal hedge or edge that transitions from the more controlled areas of the garden to the wilder areas beyond. When transitioning to a woodland, choose shade-tolerant shrubs that can thrive both in the shady woodland edge and the interior. Witch hazel (*Hamamelis* spp.), spicebush (*Lindera benzoin*), sweetshrub (*Calycanthus floridus*), or one of many viburnums will be happy in part shade and well-drained soil. If the soil is moist or even wet, try one of the hollies (*Ilex* spp., excluding the × *meserveae* hybrids) or highbush blueberry (*Vaccinium corymbosum*). Most shrubs will grow more densely at the edge of the woodland than in the interior, so keep this in mind as you determine placement. Keep the spacing uneven, avoid straight lines, and retain some openings to see into the woods, to truly integrate.

If the edge is between the garden and a meadow, shrubs can be used to as a backdrop to the garden, separating it from the meadow. Keeping the height of the shrubs in scale with the surroundings and maintaining the view over them will be important in this scenario. Choose a mounding, suckering shrub that can hold its own against the meadow and can be easily managed, either by regularly cutting out the tallest stems or by routinely cutting it to the ground every couple of years. Some shrubs to consider are red-twig or gray dogwood (*Cornus* spp.), bush-honeysuckle (*Diervilla* spp.), bayberry (*Morella caroliniensis*), meadowsweet, spiraea (*Spiraea* spp.), or one of many wild and carefree roses (*Rosa* spp.). A root barrier can protect perennials planted in front of the shrubs, but if you have a lawn, you can use your mower to keep the shrubs from advancing into it.

Shrubs can also soften the hard lines of a wall or fence. Choose climbing vines if you don't have the horizontal space for shrubs. Some vines, like climbing hydrangea, will grab on by themselves; others, like wisteria or clematis, will need to be tied off to get to the top. Clump-forming grasses can also be used in this situation. They have the benefit of taking up only modest depth, turning a lovely straw color, and offering movement throughout the winter in front of a static element like a wall.

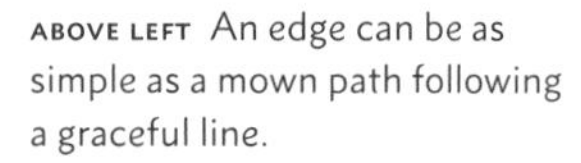

ABOVE LEFT An edge can be as simple as a mown path following a graceful line.

ABOVE RIGHT Red-twig dogwood makes a colorful transition from a lawn to a wild area.

RIGHT This garden requires no edge, thanks to naturalistic plantings and a porch that functions like a bridge to the surrounding woodland.

Edging your perennial beds with a low wattle fence, bamboo loops, or other decorative elements can define the line of the bed when perennials are dormant. Consider the ease of mowing and maintaining the edge before investing in edging material that might make your life more difficult. It can be energy-intensive to maintain a bed edged with irregularly shaped rocks next to a lawn, since grasses have a tendency to grow under and between the rocks.

Raised beds, another feature with hard edges, create relief for light to play on. Consider seeding in a cover crop for winter enjoyment or just admire the geometry of the empty beds.

The hard edge of this circular brick wall signals to visitors that they are leaving the garden as they enter woods.

TWO

PART TWO

CONTRAST

Emphasizing Polarities during the Season of Extremes

Without the unifying element of abundant green, photosynthetic leaves covering all living surfaces, the winter scene celebrates contrast. A powerful design principal, contrast is naturally at your disposal with waning winter light and bright white snow. The visual emphasis is on the arrangement of garden elements as they stand apart from the ground, so much of which is obscured or hidden in the summer.

Plants that blend into their surroundings in the summer may find themselves standing alone. Herbaceous neighbors are absent, accentuating the form of shrubs and standing foliage while contrasting with their surroundings. Thoughtfully arranging your beds for winter enjoyment provides a refreshing alternative to the one-dimensional look of a border cut back.

During the winter, we are able to see through deciduous shrubs and small trees that formerly blocked views. The contrast between foreground, middle ground, and background accentuates and makes the most of layering and heightens the perception of depth.

Arranging garden elements in the winter to emphasize differences in texture, color, form, and material will keep the garden engaging. Make the most of high contrast to add drama and excitement at a time when they are in short supply.

ABOVE LEFT Street trees cast dramatic shadows on house and garden on a brilliant day.

ABOVE RIGHT The stems of paper birch (*Betula papyrifera*) stand out when planted in front of evergreens.

LEFT An ordinary picket fence casts an extraordinary shadow when the ground is uneven.

The shadows of this hazelnut (*Corylus avellana* 'Contorta') twig wreath add another dimension to this simple decoration.

SUN / SHADOW

The shortened days of winter turn us all into sun worshipers. Bringing sunlight into the house elevates our mood and reduces the need for heating. Trimming or even removing overgrown conifers can improve your solar gain. If you rely on trees or shrubs for privacy, consider window coverings that can be opened during the day or place your privacy plantings at a distance from the window to allow light to enter. Using deciduous trees to shade the south side of the house will keep it cool in summer but will leave a clear path for winter sun.

Shadows cast by trees and garden structures stretch out on the ground, creating elongated patterns that lengthen to their extreme the winter. On the snow, those shadows can be especially graphic in a black (or blue) on white composition that changes perceptibly almost by the minute. Likewise, shadows that are cast deep into your home have a similar graphic and ephemeral quality. Deciduous trees are unparalleled as pattern makers. A well-placed tree or shrub some distance from a south window will cast intricate fractal patterns of branches that move with the wind and season.

The cedars' structure is amplified by snow, even in the shade.

To the north of the house, its deep shadow will envelop a lot more of the garden during the winter. Plants with light-colored foliage or bark will be especially appreciated here; likewise, a light-colored focal point will show up best when the surroundings are cast in shadow. A dense, deep-green stand of conifers may not be the best choice unless it is performing an essential purpose, acting as a wind or visual screen. If this is the case, planting deciduous shrubs or shade-tolerant grasses in front of them can liven up the dark shadows.

Moon shadows appear extra vivid during the clear nights of winter, casting enchanting patterns on the ground or snow during the weeks surrounding the full moon. I cherish the night view out a second-story window when I can admire the moon's luminescence and its reflection among the shadows cast by the bare branches.

Winter is a time of extremes. The sun can seem harsh and bright when reflecting off glass or snow. Shadows, however, remain dark and cold. Snow and ice linger longer in the shadows, so planting hardy trees, shrubs, or herbaceous plants is essential. Sometimes, however, having a more complete snow cover through the winter can benefit your less-hardy perennials by preventing the freeze/thaw cycles that happen regularly in sunny locations. Experiment with what works for you and the plants in your zone. Exploiting the late thaw in the spring can also extend the season of cherished bulbs. Plant them in several locations where they will bloom at different times; they will emerge later and perhaps last longer in the shadows than in a sunny location.

Shadows require a light source and an object to intercept the rays. Think about how you can employ shadow to enhance depth, dimension, and perspective in the garden. Or play with the arrangement of light (whether it be the sun, moon, or an artificial light) and a garden element to paint with shadow and create something new or unexpected.

Paper birches brighten shady spots with their white bark.

Morning light reflected by a little stream.

LEFT Crabapples catch early morning light and contrast with background shadows.
RIGHT The seed pods of silverbells (*Halesia tetraptera*) glow in winter light.

LIGHT / DARK

Winter is a dark time. As fall progresses, daylight hours diminish until we are eating both breakfast and dinner in the dark. The long nights make me revel in the few hours of sunlight each day—and what light it is! The golden, sideways light of winter is something I look forward to all year. And unlike in midsummer, sunrise happens at a time when I am likely to be looking out a window or walking out the door.

When the azimuth of the sun is low, prolonged sunrise and sunset make for garden drama. Take advantage of the sunrise by placing a tree or ornament in a spot that will be illuminated by the rosy, lemon light of early morning. Experiment with planting a small deciduous tree or shrub in the southeast, where the first morning rays of winter will shine through the branches. The backlighting will highlight the edges of the plant, accentuating the beauty of its form and structure. Imagine the sunrise illuminating a dusting of snow or the sparkle of ice crystals on its branches. Use your old Christmas tree or a fallen branch to find the right location, then mark the spot for planting in the spring.

LEFT Seed heads stand out for winter enjoyment.

BELOW A frosty scene can be lovely even on a cloudy day.

OPPOSITE On the shady side of a house, this urn really shows off against the snow cover and background of conifers.

OVERLEAF Sunset and twilight can be better appreciated in winter, when the moon is higher in the sky.

Place a sculpture or other garden element toward the west, where sunrise will illuminate the face of the object. A light-colored object placed in front of a deep-green conifer will really stand out in the light; consider a clump of ornamental grasses or a beautiful rock and shrub planting.

The light-against-dark motif can be put to work at all times of the day. The structure of a leafless shrub with light-gray bark like a winterberry holly or witch hazel can really show off with a conifer or a dark-colored fence or wall as a backdrop. Likewise, a shrub or tree with dark bark will be much more attractive in front of a light background. Snow makes the brightest background for a deep-green conifer, a sweet birch tree (*Betula lenta*), or the darkened seed heads of perennials such as echinacea or black-eyed Susan. Placing conifers and ornaments with solid forms in the foreground will highlight their outlines, while the structure of perforated forms like leafless shrubs or trees will be emphasized. Pruning can improve both the outline and structure of conifers and deciduous trees and shrubs.

Early morning light enhances contrast, especially with the cover of snow.

Sunset happens discouragingly early in the winter. Instead of dreading it, create a scene that celebrates every last winter ray. You can employ the same techniques discussed for sunrise to enhance the effects of sunset in your garden.

Stonehenge and other monuments to the winter solstice mark with stone the sunrise and sunset on the shortest day of the year. Observing the passing of the solstice brings to mind that although winter is just getting started, the amount of daylight will begin increasing day by day. If you locate a rock, sculpture, or plant to intercept the rays of sunrise or sunset on the day of the solstice, you will be reminded of your connection to your garden as well as your place on Earth and Earth's place in the solar system. Reinforcing your partnership with the sun in wintertime feels better than resenting the lack of daylight.

CONCEALED / REVEALED

The Japanese language has an expression for the moment when snow clumps on the bare branches of a deciduous tree or shrub: *hatsu hana*, literally "first flowering." This is a euphemism, but it reminds us to appreciate the beauty we see every day. (Rest assured, this expression is also used for the first actual blossoms.)

Given changing climate patterns, it is hard to know what future winters may bring. Creating a garden that will thrive and look attractive in all possible conditions may seem overwhelming, but good design and appropriate maintenance will go a long way toward achieving a pleasing garden, no matter what comes our way.

While growth and change slow in the garden during winter, the effects of weather do not. Fluctuating climatic conditions are the norm here in coastal Maine. Unlike some inland areas at our latitude, we do not usually have snow cover for the entire winter. Enduring alternating snow, rain, freezing, and thawing, our gardens and our plants need to be especially well equipped to deal with change—and so do we.

Some gardeners in coastal Maine mulch their perennial beds with fir boughs. If the boughs are clipped after a hard frost, they will retain their needles throughout the season, unlike those of most other conifers, which turn brown and drop their needles after a couple of weeks. A layer of boughs confers a degree of insulation to the roots of dormant plants when the ground is bare and temperatures plummet. The ground also stays cooler during warm spells that might send the wrong signal to half-hardy plants.

If you have done some planting in the fall and are worried about your new plants heaving out of the ground during repeated freeze/thaw cycles, fir boughs are your answer. Not only will they protect your new plantings, but they look wonderful, whether or not you get a nice covering of snow. You don't live where fir boughs are available? While not as attractive and green, a thick mulch of leaves or straw will accomplish the same thing. However, you may need to weigh these materials down with limbs to prevent them from blowing. And don't forget to remove your mulch when the ground elsewhere has thawed and spring is really on its way, especially if you have ground covers in the area.

I have learned the hard way that wise gardeners wait to cover the beds until the ground has frozen, at least at the surface. By this time, local rodents have made a home elsewhere and will not be as likely to move into the mulch hotel you have just created with leaves, straw, or boughs. Meadow voles, who live in all the northern states east of the Rockies, and mice, who live everywhere, can be a nuisance to your woody shrubs and young trees during prolonged snow cover. You will not see the damage they are doing to your plants as they move through tunnels under the snow, nibbling on bark and girdling or even killing your woody plants. I have seen a nasty job done on juniper, lavender, and young fruit trees of every variety. You can easily protect trees with plastic or wire mesh trunk protectors that can be removed during the warmer months. Sprawling plants like junipers are more difficult to protect, but bird netting stapled to the ground can deter deer and rabbits from nibbling and is not that offensive to your human visitors.

Snow can put your plants in danger by weighing down the limbs of conifers or deciduous plants before leaf fall. Most of the time, there is nothing you can do to prevent this type of damage. Wrapping conifers in burlap can protect living needles from drying out in harsh winds and can prevent multiple trunks from splaying out during a sticky snow event. Northern white cedars (*Thuja occidentalis*) are especially vulnerable to this.

OPPOSITE, LEFT *Hatsu hana*—the first spring "flowers" of snow—collecting on a magnolia, the buds of which are swollen throughout the winter.

OPPOSITE, CENTER Fir bough mulch is decorated with winterberry (*Ilex verticillata*).

OPPOSITE, RIGHT Icy conditions highlight red-twig dogwood (*Cornus sericea*).

RIGHT Snow highlights garden structures as well as plants left standing for winter.

Also be aware of snow sliding off a roof. If you are looking for a reason to get rid of those overgrown "foundation plantings," this is it! Instead of breaking out the plywood sandwich boards to cover shrubs in danger of plummeting snow and icicles, consider just moving them to a safer position in the garden and replacing them with herbaceous perennials or ferns that will be dormant during the barrage. Other options are woodies that can take being cut back to the ground annually if they get clobbered, such as hydrangeas of the *arborescens* type ('Annabelle', 'Incrediball', et cetera), flowering raspberry (*Rubus odoratus*), or bush honeysuckle (*Diervilla lonicera*).

LEFT An early morning frost on winterberry.

BELOW Flower buds of azalea (*Rhododendron* sp.) emerge from the snow.

OPPOSITE A uniform row of blue beech (*Carpinus caroliniana*) provides plenty of visual stimulation, especially in the snow.

The benefits of snow go beyond its function as a great insulator (particularly for plants whose crowns are near the surface). If you are lucky to live far enough north to get a good snow cover in the winter, it may actually fertilize your garden. When the snow melts and the ground thaws, growing plants can access essential nutrients such as nitrogen and sulfur that snowflakes picked up in the atmosphere and delivered to your garden. Friendly microorganisms in the soil are required to make these nutrients available, so combined with organic practices, snow can give your garden a boost. The upside of freezing ground—with or without snow—is that when ice particles form and expand in the soil, they help loosen the soil and counteract compaction. Understanding the dynamics of freezing and thawing can help you devise strategies to keep your plants thriving in uncertain times.

UNIFORMITY / DIVERSITY

In the fall, when the white pines surrounding my house shed a portion of their oldest needles, a yellow (soon to turn gold) layer veils the plants and ground. I look forward to this moment of transformation, when the needle curtain creates a new garden topography without the distraction of individual plants. While I cherish this moment of uniformity, I am just as glad it doesn't last. The needles eventually find their way to the ground and turn russet, a color that carpets the floor of the garden and becomes a backdrop for the changing colors of deciduous shrubs.

Snow blanketing the ground presents the ultimate in uniformity. The textures and colors of low plants and ground covers are reduced to a clean white. This reduction in visible detail creates an austere and pleasing scene;

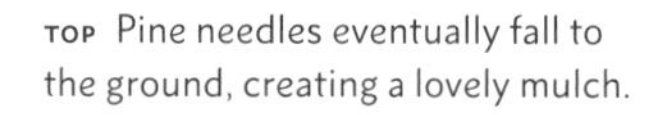

TOP Pine needles eventually fall to the ground, creating a lovely mulch.

CENTER LEFT AND RIGHT Mounds of heather repeat the shapes of boulders, adding color and texture to this composition.

BOTTOM RIGHT Similar forms of rhododendron, juniper, and heather combine to create an intricate tapestry.

A blueberry field is punctuated by randomly placed boulders.

the ephemeral nature of this transformation helps us to see things in a new light. An attractive background or focal point can make the scene especially lovely.

A balance between uniformity and diversity creates harmony. The fallen needles and the snow provide a temporal uniformity in an otherwise diverse garden. In spatial terms, a harmonious balance is apparent in a blueberry barren punctuated by boulders or the view of an island surrounded by water.

A powerful sculpture echoes the colors and shapes of its surroundings.

LEFT Rows of pots are transformed into a regular but intricate three-dimensional surface by a blanket of snow. RIGHT A line of cedars (*Thuja occidentalis*) contrasts with a uniform but undulating lawn and woods in the background.

We can be intentional in how we employ the contrast between uniformity and diversity, beyond merely waiting for Mother Nature to drop needles or snow. Think about the contrast between lawn and garden bed or between ground cover and shrub. How can that contrast be maintained or accentuated through the winter, when perennials are dormant and bulbs have yet to emerge? Adding conifers and deciduous shrubs to a garden bed is one way. An urn or large rock can also break up a uniform horizontal plane, as will small trees such as crabapple (*Malus* spp.) or witch hazel (*Hamamelis* spp.). Changes in topography—uneven ground or bedrock emerging from the horizontal—affect the way light plays on the ground and add a degree of diversity to a scene.

We need contrast to appreciate uniformity, just as an empty space is defined by the fullness surrounding it. Direct attention by punctuating the view with forms that draw the eye from one place to another. Add changing light and weather conditions, and you will have created a sight that stands the test of time, penetrating our busy lives by calling our notice outdoors to reconnect with nature.

I am always in favor of ecological diversity, especially as it encompasses our native plants and animals. On an aesthetic level, however, diversity can easily be overdone. When incorporating novel forms into the garden, do so sparingly. Remember the way unique elements tend to stand out in the winter and be conscious of the line between enriching and overwhelming. Repeating shapes or colors can create meaningful and engaging rhythm throughout the garden.

After the exuberance of summer, winter will always seem stark. Avoid incorporating too much color and diversity into the winter garden; instead, strive for a balance that provides restful views that incorporate elements or qualities that surprise and hold the attention.

White sand in the foreground could not be confused with a natural formation, as it contrasts with layers of surrounding plantings.

Framing a view of an uncultivated forest.

WILDNESS / CULTIVATION

Have you ever heard someone say they are "cleaning up the woods" or trying to make a new planting "look natural"? These two diametrically opposed desires are often expressed by the same individual—sometimes myself! Why do we tidy up what nature has provided, yet, at the same time, desire landscapes that look as though they have transpired on their own? I think it is because we see the beauty in nature but get overwhelmed by the details. We think we can understand the simplified version of nature that we cultivate in our gardens if it appears orderly and visually comprehensible. But could we learn to appreciate and live with both cultivation and wildness?

You can create a stark juxtaposition in your garden by imposing an overtly human-made (or highly controlled) object into a wild environment or by inserting a wild object into a human-made environment. Place

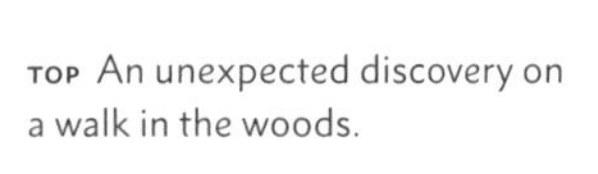

TOP An unexpected discovery on a walk in the woods.

BOTTOM LEFT An opening in the fence frames a view of uncultivated woodland, creating a pleasing composition.

BOTTOM RIGHT Stumbled upon in the wilderness, this simple sign may cause an explorer to pause and smile.

Natural materials arranged in a distinctly human way can surprise and delight an unknowing wanderer.

a sculpture in the woods or a wild planting in an urn, create a living wall or roof, or experiment with something as simple as a piece of driftwood in a perennial garden; scenes like these highlight the tension between nature and humans.

The British artist Andy Goldsworthy is a master of this technique; his worldwide acclaim speaks to its universal appeal. He arranges natural objects such as leaves, twigs, or stones in a distinctly human way. His works are both ephemeral and permanent—he considers the ephemeral works to be intuitive explorations informing the permanent work. Goldsworthy says, "The ephemeral is the breathing in, the nourishment, whereas the permanent is the breathing out." Why not try creating your own ephemeral works of art in the garden? Let this practice bring you closer to your

A restful scene contrasts with the winter view of a distant mountain.

landscape, its inhabitants, its changing light, and its weather. Maybe these works, too, will lead to something permanent.

Consider a Zen garden—the iconic version, with raked gravel undulating around a few substantial rocks, the only green is the moss growing at their bases. The austerity of this pared-down representation of nature is valued by those hoping to glimpse an inner truth through the practice of meditation. The sparse landscape is specifically designed to provoke contemplation about what is left out, rather than about what is included.

All of us can appreciate the beauty of these gardens (regardless of whether we seek enlightenment) and adapt their principles to our own spaces. The highly abstract gravel, carefully raked into simple patterns, represents water and its changing nature, while the jagged rocks created by geological processes over the past millions of years represent mountains

A sculpture reflects the spirit and tranquility of the coastal Maine woods: seeing is feeling.

and permanence. This contrast is what makes Zen gardens work for our modern sensibilities.

Stark contrasts are good when you want to make a point, but how can you embrace both contrast and harmony? Easing the transition from a human-made environment to the wildness beyond can be a challenge. In a Zen garden, the transition is usually abrupt, marked with a wall or fence that clearly sets the space apart from the reality outside. This is not always practical or desirable in a home garden. Your home and any hardscaping associated with it are obviously human-made; in the winter, these are the places from which you will be viewing your landscape most often. You may want to create a transition or a bridge to the wilder areas beyond the confines of your house or garden. Framing a view with a gate, a hedge, or an allée can enhance the connection and focus the attention. Transitioning from the garden to the woods with shrubs that thrive at woodland edges in your area can provide a bridge. Making the transition with an undulating line rather than a straight one can also help.

If everything in your garden is clearly affected by the human hand, it may fail to be perceived as a garden—it might be considered sculpture or architecture instead. Conversely, if there are no perceptible signs of human involvement, it will be wilderness, not a garden. The amount of intervention each gardener or designer exercises should depend on the needs of both the ecosystem and the beings who inhabit it. Whose needs and which needs should take precedence? What do we want from our gardens—comfort, humor, connection, humility? What does the ecosystem require? When we have so much yet to learn about the multitudes of organisms that occupy even the smallest garden plot, erring on the wild side might not be a bad idea.

MACRO / MICRO

Winter elicits so many changes in the garden. For us northerners, leaf fall is probably the most notable. As I drive to and from work each day in the autumn, I notice so much—my view extends into the distance, far beyond the first line of trees and shrubs at the roadside. Winter can be expansive in that way. Instead of mourning the fallen leaves, view them as a curtain opening to reveal something more, something distant, something different.

The ha-ha, which is a retaining wall set below grade, was invented in the eighteenth century to obscure the boundary between the garden and what lies beyond. This idea is most often employed today in the infinity-edge swimming pool. While the ha-ha was frequently used to prevent the view of distant fields from being marred by fences to keep out roaming livestock, it was also a way to visually connect the foreground of the garden with the view in the distance. Consider linking your garden to a distant view, whether it's a part of your garden or not. You can do this with a ha-ha or by echoing distant plants, shapes, or other features.

Give yourself something to look forward to by capitalizing on views that open up only in the wintertime. Intentionally frame winter views of

Trees become transparent after leaf fall, opening up winter views.

A distant winter view framed by a veranda.

A bench in this wide-open view gives the scene context
and reminds us to linger and appreciate.

LEFT Taking in the long view.

BELOW LEFT A close-up of colorful heather (*Erica* sp.).

BELOW RIGHT Junipers and heathers flank this path, providing an opportunity for intimate appreciation.

A sculpture frames an archway of beech (*Fagus sylvatica* 'Dawyck Gold') that leads the eye across the pond to another sculpture on the opposite bank.

LEFT Evergreen ground covers like this bearberry (*Arctostaphylos uva-ursi*) can be appreciated up close. CENTER The golden branches of this Japanese maple encourage examination when viewed from a nearby bench. RIGHT A vine's intricate, twisted trunk provides textural interest.

attractive features such as a beautifully pruned tree, a glimpse of a stream, or a distant prospect. A fence or trellis can be used to make a frame, as we will discuss later—or try a small tree or a pair of trees or shrubs. While we humans tend to like wide-open spaces—this is probably baked into our survival instinct—a more limited window, whether spatial or temporal, may be more treasured.

Conversely, following leaf fall, some not-so-attractive views may present themselves. These situations should be prioritized for action the following summer; I don't want to spend my winter dwelling on something I could have moved or screened rather than enjoying everything else the garden has to offer. Would a fence be appropriate? A hedge, some twiggy shrubs, a conifer, maybe even an artfully constructed brush pile? Blocking out unwanted views can enhance just about everything else.

When the weather outside is threatening—snowing, raining, or just misty—our view tends to close in, bringing the garden boundary forward. This close-up perspective evokes feelings of security and comfort for all but the most claustrophobic among us. This partitioning off of a piece of nature to control and to appreciate its details close-up is embedded in the meaning and practice of gardening; in fact, the roots of the word are thought to have meant "enclosure." Create winter motifs that celebrate the near—an intimate view out a window or a ground cover near a path that would go unnoticed in the summer. Strategically placed plants or garden structures can define, limit, structure, and showcase the proximate elements of your garden while preserving the mystery of what lies beyond.

Create a composition that you can admire as you pass, by adorning the paths you regularly take with plants that have attractive winter foliage or texture. Consider plants that will look great at night under pathway lighting, such as mosses, bearberry (*Arctostaphylos uva-ursi*), or the many available evergreen ground covers.

While the gentle lighting of a dreary day creates a somber mood, clear days and nights exhilarate and energize. That bluer-than-blue sky can be magical when viewed above the treetops or as a backdrop to your dormant garden. Every hint of color is magnified in winter light. Snow on the ground can bring the colors up another notch. Distant views will sparkle through the clear winter air. You may be able to see further into the distance than at any other time of the year. Make it your goal to create a winter view that you will find delightful and uplifting, regardless of the weather.

LEFT Bright yellow lichen lives happily on this stone wall. CENTER Light snow reveals the geometric structure of this spruce. RIGHT Some apples (mainly crabapples) remain on trees well into the winter, to be appreciated by humans or visiting wildlife.

THREE

PART THREE

EMBELLISH

Adorning the Garden for Year-Round Enjoyment

I came to gardening through a love of plants—specifically, my love for the plants in the native landscape. Coming across my favorites in the woods makes me feel like I am greeting old friends! For me, the plants are the living stars of the garden; adornment is secondary. However, when the plants are at less-than-peak performance, other elements in the garden can take up the slack by creating focal points, directing attention, and making the garden shine.

Some embellishments, such as fences, trellises, and stone walls, are functional throughout the year. Stone in the garden has the advantage of looking completely natural, often as if it had been there all along. It is the only material we use to construct our gardens that is literally millions of years old!

Other embellishments, including those that are undeniably human-made, can add color or texture, whimsy or humor. As with so many garden elements, any ornament will be especially noticeable in the winter. Often, they are most effective if they are less noticeable in the summer, when the plants are having their moment, in order to save their impact for the winter months. Seasonal ornaments can offer something special to look forward to during the darker months.

A large fixture casts light on a surrounding cedar tree.

Low-voltage lighting fixtures come in a range of styles.

LIGHTING

A clear night sky, the kind we most often have in the winter, is an inspirational sight. If you live away from a city or source of bright night light, you will likely be able to admire the Milky Way and the winter constellations. And don't forget moonlight, especially on the snow—it can be simply brilliant. Natural light in the winter is a scarce commodity that increases our appreciation—which must be why winter celebrations so often include the element of light. However, we can't rely on natural light during the long winter nights for practical purposes such as illuminating the walkway to the house's entry or the way to and from the car.

Rechargeable solar lights are a good solution for areas that you don't mind lighting for several hours every night. Be aware that the period for which they emit light each night shortens, as diminished temperature and reduced daylight hours affect the amount of recharge. The result is fewer hours of illumination just when you are needing it more. In addition, the batteries have a relatively short life, so these lights are not effective as a permanent solution.

Low-voltage exterior lights are a larger investment, but you can turn them on and off at will, set them on a timer, and dim them. Linked to a motion sensor, they will turn on only when needed. With countless

Solar landscape lighting is used to dramatic effect.

LEFT An arrangement of greens and tree prunings is lit for indoor or outdoor decoration.

BELOW Candlelight in the garden can be magical.

styles available, these are a good option for flexible permanent lighting. You might use rechargeable solar lights to try out options for positioning low-voltage fixtures before committing to a long-term solution.

Balance your outdoor lighting needs with the needs of wildlife and neighbors by following a few basic rules:

- Use only as much lighting as necessary. Consider the location, number of fixtures, brightness, and angle of each lamp.
- Make sure lighting is directed downward toward the walkway or path, not at the sky or your neighbor's yard (known as light trespass).
- Use motion sensors with timers to light up areas only when they are in use, for a limited amount of time.
- Make sure the color temperature of the light is below 3,000 kelvins (avoid light at the blue end of the spectrum).

FAR LEFT Winter color in the garden can be as simple as a coat of paint.

LEFT A rain chain frozen in action sparkles in the sun.

BELOW A spot of color in the landscape.

Consult the International Dark-Sky Association for recommendations and resources for low-impact outdoor lighting.

In addition to fulfilling your practical needs, lighting can also add drama to the garden. Uplights can be used at the base of trees that you would like to highlight at night, as long as they are focused and do not wash the sky with light. Downlights mounted in a tree or on a nearby wall are even better. Spotlights can be focused on a sculpture or decorative element. Consider putting lights on a timer or manual switch to control how often and how long they will be on and to make sure the bulbs cast a soft light.

Strings of lights can be festive when used to highlight a tree, shrub, or structure. As with any focal point, limiting the area covered by lights and opting for quality rather than quantity will define the center of attention. Too much of a good thing encourages the eye to wander and dilutes the effect.

Don't forget about candles. With attention to safety, there is no reason they can't be used outdoors. I'm sure you've seen DIY instructions for making luminaries out of paper bags, mason jars, or blocks of ice. For a special occasion, these have just the right glow and homemade touch to create a cozy and welcoming mood.

On other nights, find opportunities to cherish the clear winter night sky—it can be breathtaking! (Or is that just the cold?)

ORNAMENT

Garden ornament can run the gamut from a found item to a finely crafted bench or work of art. When placing a garden ornament, we should be especially conscious about what we are choosing (and why) and where it will be most valued and resonant. Can it perform a function that will be especially appreciated in the winter or solve a specific winter garden problem such as redirecting attention from an unwanted view? Do remember that most objects in the garden are much more visible during the winter than when foliage is lush and abundant.

Think about which ornament would complement or enhance your garden. Maybe you own some unique object, and you "just like the way it

Large-scale sculpture and garden art provoke thought at every turn in this twenty-acre garden.

looks." Maybe you have created something yourself or have found something that speaks to you. Does it evoke a feeling or a memory or provoke some thought? Ask yourself why you like it and how it could fit in your garden space. Of course, you also will need to consider if it is resistant to weather, or will it be an experiment in impermanence? And don't forget about color. A splash might turn an ordinary fence, door, or bench into an ornament that will brighten a dull winter day.

Ornaments that help us perceive the invisible or remind us that nature is at work are very well suited to the garden winter. An ornament that changes with the weather, moves with the wind, or makes a faint noise can bring you closer to your garden, even if you don't get outside much.

A rain chain can be considered an ornament, even though it is a functional device that conveys water from the roof to the ground. Used in place of a downspout, it can be simple or decorative, guiding water to a rain barrel, surface drainage, or an underground pipe. Designs range widely from basic links to ornate cups that funnel water downward. The beauty of a rain chain is that it makes a rainy day eventful and creates a temporary water feature in your garden, but be aware that you might need to remove them in the deepest of winter to avoid ice accumulation. Taking small opportunities like viewing a rain chain can reinforce your connection to nature and your garden and uplift you in all seasons.

Think about how an ornament will function: Will it provide a contrast between the natural and the human-made? Create mystery? Add a spot

Colorful garden sculpture is most appreciated on a monochromatic day.

CLOCKWISE FROM TOP LEFT This sculptural trio feels at home in a woodland setting.

Organic plant forms embellish an entry gate.

A multitiered stone lantern contrasts with the natural and cut stone in this garden.

A whimsical sculpture is adorned for the season.

OVERLEAF The grounds and gardens of the John Hay Estate at The Fells on Lake Sunapee in Newbury, New Hampshire, are full of year-round inspiration and delight for visitors.

of color to relieve a monochromatic winter? Do you need a focal point to draw attention or to liven up a dull corner? What can be used to slow down a visitor or entice someone to explore? In Persian, Chinese, and Japanese cultures, many garden objects have symbolic meanings based on their shape, material, or metaphorical significance. A lantern might symbolize enlightenment as light is a metaphor for wisdom, or the shape of a particular stone may evoke the shape of a seated Buddha or may conjure feelings of strength and power conferred by a mountain. Embodied meaning and purpose enhance any garden ornament.

Think about where to place your ornament. Finding a suitable location can be tricky. Will it be viewed from the house, stumbled upon in the woods, or used to compliment a planting? Does the shape of it relate to something in the garden? What about the color, texture, or material? Are you using it to contrast with the surroundings, to blend in, or to create a bridge or transition between two different areas? Check that its scale is appropriate. If your ornament looks too small for its space, consider adding diminutive plantings that grade into larger plantings in the surrounding. If it looks too big, try mitigating with larger-scale plantings such as shrubs or small trees that will help it to integrate.

So many questions! If you can come up with answers, they will guide you in selecting ornament that looks appropriate and intentional. As with so many elements in the garden, ornament is easy to overdo. Using restraint can prevent a garden that squanders attention and, instead, directs it to the places you most want it to linger.

To get the placement right, use a plastic pot or something of similar size and shape to simulate the ornament during the winter and mark the location for spring digging. It is surprising how forceful a bit of frost can be. A good base with adequate drainage is necessary to prevent top-heavy items like birdbaths from listing or even toppling. To accomplish this, work when the ground is not frozen. If the item is fragile, it will be well worth your while to dig down a foot or so. Add crushed rock for drainage and then cover with filter fabric and soil or a flat stone before installing the ornament.

A fence and trellis combine with a stone wall to surround this formal garden.

STRUCTURES

Some gardens are designed to provide a foreground to the view beyond the garden, while others are meant to obscure the view of the outside world and create a refuge within. Many of us in the north can have it both ways, looking inward in the summer and—as leaves fall and views open up—outward in the winter. But sometimes a more permanent year-round structure is needed to delineate one area from another, provide support for growing plants, or offer a feeling of safety or sanctuary.

Fences

Installing a fence can be expensive, and figuring out exactly where to put it can be a big decision if there is not a perfectly obvious location such as your property line. Experiment by marking the area where you think you might want a fence by sticking bamboo stakes or brush from pruning into the ground; this will give you an idea of how the fence line might look and whether it will accomplish your goals.

The purpose of the fence will help you decide its dimensions, materials, and construction. Will it be open or solid, straight or curved, made out of wood planks, pickets, bentwood limbs, wattles, wire mesh, or composite materials? Rustic fences can more easily define curves, with milled wood panels or pickets more suited to straight lines. The permanence of the fence and how will it relate to the architecture of your house or other garden elements should also be considered when selecting materials.

LEFT It is easy to form curves with a wattle fence made from spirea brush.
RIGHT This openwork fence is high enough to keep deer out without visually walling off the garden.

LEFT An attractive fence keeps out deer while supporting climbing hydrangea (*Hydrangea anomala* ssp. *petiolaris*). RIGHT Ironwork botanical designs hint at the garden behind this elegant wooden gate.

Gates

Gates require a moment of pause. They can be functional, decorative, or symbolic. They may serve as a welcome sign to enter or they can be a barrier, visual and/or physical. As with fences, the purpose of the gate will help you decide its design, as well as how it will relate to other structures such as the fence or trellis of which it is a part. For some purely functional purposes like keeping deer out, the gate or a lentil above must be 7 to 8 feet (2.1 to 2.4 meters) high; to deter rabbits, it must nearly reach the ground. If you want to welcome human visitors, however, a more open gate with view of the garden beyond is most inviting.

Trellis, arbor, gazebo, pergola

A trellis or arbor can be a beautiful winter feature, even when devoid of foliage. It can be used to frame a view beyond or it can signal the entrance to a special part of the garden. An arbor can support perennial climbing vines such as grapes, roses, or clematis or annuals such as wild cucumber or pole beans. Considering leaving obelisk-shaped trellises standing in your vegetable garden in the winter, just to catch the snow or your eye.

Larger structures such as a gazebos and pergolas define a space in which to linger. While a gazebo may be strictly for us humans, pergolas are

This rustic trellis marks the entry to a vegetable garden
and frames the view to a toolshed.

This grape arbor is a lovely place to linger on a summer night
or to spend a moment to take in the winter view.

The path through this garden is framed by structures at both ends that focus the visitor's view.

frequently covered with plants. The view of the structure and the view from the structure both need to be considered when deciding where to place it. Does the location make sense in the winter as well as the summer? I have often admired the wall-less "summerhouses" used by rusticators (city folk who escaped to the countryside in the early twentieth century) and wonder what a "winterhouse" would look like.

Vista may be a grand word if you have a small lot, but you can create a long view from your kitchen or bedroom window, even if there is no water or mountain to relish by cultivating a single line of sight to a distant feature. A device such as an opening in a fence or a trellis can be used to frame a lovely tree, or an opening in the canopy that receives a ray of sunshine. You can accentuate the distance by creating a wider opening at the

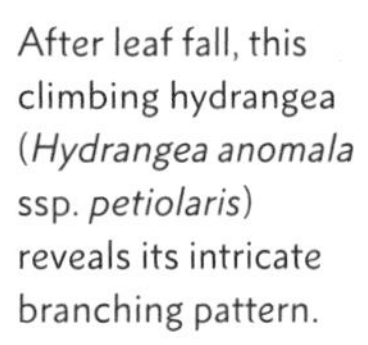

After leaf fall, this climbing hydrangea (*Hydrangea anomala* ssp. *petiolaris*) reveals its intricate branching pattern.

end from which the vista will be viewed and narrowing it as it reaches the destination. By framing the vista, you prevent the eye from wandering and give it a place to rest. Vistas can be constructed throughout the garden, but concentrate on those that you can see from the house or from the areas you frequent during the winter.

Walls

If you have a large, blank wall that needs some embellishment, consider using plants to add depth. A windowless garage wall doesn't need to be boring: an espaliered tree or shrub can cloak the wall in vegetation. Even easier, a climbing hydrangea will grab on by itself. Trellis material attached to a blank wall can be a decorative element with or without vegetation, transforming an unremarkable surface into an artwork of shadow and texture.

Some plants with a winter presence that are great for training onto a fence, trellis, or wall include fruit trees, such as apple, crab apple, pear, and peach, and vines, such as climbing hydrangea (*Hydrangea anomala* ssp. *petiolaris*), Virginia creeper (*Parthenocissus quinquefolia*), and our native honeysuckle (*Lonicera sempervirens*). Some shrubs that can enliven a structure include cotoneaster, flowering quince (*Chaenomeles* spp.), and grapes. The combination of a human-made structure and a living veil of plants exemplifies the gardener/nature collaboration.

STONE

Nature has a way of placing stone in the landscape; while not random, it may appear that way to those of us who have not studied geology in depth. Coming across a house-size, lichen-encrusted boulder in the woods always feels magical and gives me a reason to stop, touch, and admire.

Functional stone is common in the garden to create stepping stones, walls, and patios. In Asia, there is a long history of using stone for decorative purposes. Early Chinese gardeners attempted to recreate the larger landscape by using stones and their arrangements to evoke images of mountain ranges and peaks. Fantastical weathered and water-beaten stones were used as focal points in the garden. In Japanese gardens, stones were valued in an animistic way, as embodiments of spirit capable of directing or moderating energy. Some stones symbolize specific mountains that hold religious significance; others have a shape reminiscent of a human-made object such as a boat or a bridge. The sculptural qualities of individual stones are valued as well.

LEFT A field stone wall mortared in place surrounds a shady garden.
RIGHT Natural stone steps are in scale with the path through this woodland garden.

Constructed on a bedrock outcropping, this naturalistic garden provides both challenges and opportunities for the gardener.

If you are lucky enough to have a large boulder or exposed bedrock in your garden, these will have to remain where the geologic forces have left them. I do like to be reminded now and then that some things are beyond my own control; rather than resenting their placement, design the garden around these features. A rock garden or scree garden can accentuate exposed bedrock. Many rock garden plants have attractive winter color, particularly heaths, heathers, and sedums.

In contrast, smaller boulders and stones can be moved into positions where they can take on a sculptural aesthetic. As with any focal point, they are best used sparingly so as not to overwhelm. A large stone can create its

ABOVE LEFT Rounded cobbles are used as a ground cover, echoing the shape of the clipped spirea.

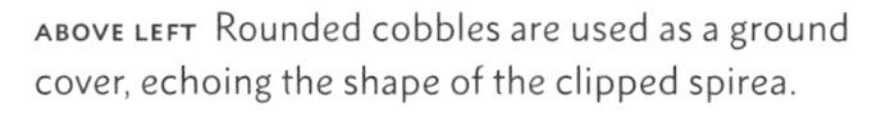

ABOVE RIGHT River stones flowing toward the base of a rain chain are in harmony with a stone foundation and column bases.

RIGHT Stone is used to edge this pool, which incorporates a natural stone boulder and sculpted stone lantern.

own microclimate by heating up in the sun and retaining cold longer than its surroundings. In the winter, these properties can be used to your advantage by understanding that dynamic and planting accordingly.

Smaller stones used to evoke flowing water can be an effective garden feature. A dry streambed that functions when drainage is needed is practical as well as symbolic. Raked sand or gravel is used to evoke water in Japanese dry gardens; the small grains even flow like water.

A favorite stone can be used as an ornament in a spot that you often pass. A pair of stones can be used as a gateway or a sentinel at an entrance. If you want a natural look, stones in the garden must be "planted." By this

LEFT Natural stone contrasts with precise stone steps and walls.

OPPOSITE Iconic fieldstone defines the edges of an old roadbed.

I mean that they should be partially buried or at least look like they are protruding from beneath the soil's surface. The only place in nature where you will see a stone sitting on top of the soil is where the soil has been eroded beneath the stone, as in a riverbed or if the stone is lying beneath a cliff from which it has been only recently cleaved off by ice action. When planting stones, create a base of well-drained material like crushed rock to prevent the stone from heaving or tipping over.

Plantings around stones can help them appear to be more settled and integrated into their surroundings. Grouping stones yields greater sculptural effect. Don't worry about rules that prescribe an odd number of stones—just use your intuition and critical aesthetic to decide what looks right. Attention to spaces between the stones and future plantings will guide your placement. You can simulate stones with buckets or bags filled with newspaper to represent the volume of the stones you plan to position. Check their placement from windows and other viewing locations before doing the hard work of moving the actual stones into place.

A rock or masonry wall can be a strong design element, useful for dividing up spaces, creating intimacy, retaining soil, and terracing slopes. Many styles are possible, from the rubble walls that line our landscape in the northeast to those built from cut or fractured stone such as granite or shale. Each construction method has a different effect, from casual and informal to crisp and precise. A good foundation is essential for any new wall; professional help is advised in the north, where freezing and thawing take a toll.

As always, use the natural surroundings as your guide. Employing local materials and honoring local wall building traditions is probably the best way to make sure that stone structures feel appropriate and that they contribute to the sense of place.

baskets
Soap
ONLY

FOUR

PART FOUR

CARE

Creating a Big Impact through Small Acts

When I hear someone express the desire for a "no-maintenance garden," I have to smile. As the saying goes, nature abhors a vacuum, and any open ground will soon be colonized by something you probably don't want. We gardeners smother bare ground with loads of bark mulch. Because mulch contains compounds that trees produce to protect themselves from intruders, it generally does a respectable job at suppressing weeds, but it looks completely artificial. Is there another way?

I am here to tell you there is! The key is to design a garden you want to be in, want to care for, and want to admire throughout the year. By using layered plantings, you can occupy empty spaces with plants you like and reduce the need for so much weeding and so much mulch. There will always be garden maintenance, but the good news is that making your garden more enjoyable throughout the winter is mostly a matter of doing less, not more.

By finding ways to evolve or change completely the things you are unhappy with, you can avoid maintenance that takes an overwhelming amount of effort. Mother Nature does not need us to care for her plants. With some forethought and a modicum of intervention—maybe some planting next year, a little pruning and reining in—your garden can be the perfect home for any of the hundreds of plants that love living in the north and look fabulous doing it.

ABOVE LEFT When glazed pots are emptied for the season, they become ornament.

ABOVE RIGHT Full of color and texture, this garden is beautiful all winter.

LEFT Sedums like 'Autumn Joy' have sturdy stems and pair beautifully with black-eyed Susan and lavender all winter long.

OPPOSITE A garden of hardy plants does not need winter protection.

MAINTENANCE

As winter approaches, conversations turn to what everyone is doing to "put their garden to bed" before the snow flies. As gardeners, we have been trained to rake leaves, cut back perennials, and tidy up "for winter." In reality, winter may not care what happens in the garden. We should be thinking about how to prepare the garden for our own enjoyment and to make the most of the habitat our gardens provide for our nonhuman neighbors.

Cutting back

Because many insects—mostly beneficial and food for birds—overwinter in our gardens, we should leave as much standing foliage as we can throughout the winter. Yes, cutting it all back may be simpler in the fall, but waiting until spring has a couple of advantages.

You will have less to cut back in the spring, as foliage will have dried up and decayed in place under the snow. That means that as much nutrition as possible has made its way back to the plant, the soil, and the organisms who break down plant fibers into nutrients that can be reused by your

plants next season. It is kind of like mulching in place—less work for you and more food for everyone else. A few plants that I do not recommend leaving are peonies or any plants that have had fungal diseases, such as powdery mildew. It is best to cut and dispose of these off-site.

Keeping perennials with sturdy, hollow stems standing provides habitat for overwintering insects. Cut the plants that are likely to crash during the winter halfway down or bundle up the cut stems and lay them in an out-of-the-way place in your garden, where insects can still access them. Most of these insects benefit your garden by pollinating your plants, eating aphids, or becoming food for birds. Perennials with hollow stems include the milkweeds (*Asclepias* spp.), Joe Pye weed (*Eupatorium* and *Eutrochium* spp.), hostas, echinaceas, and monardas. Many grasses have hollow stems, such as switchgrass (*Panicum* spp.) and little and big bluestem (*Schizachyrium scoparium* and *Andropogon gerardii*). Shrubs like hydrangea and elderberry (*Sambucus* spp.) can be harvested to make bee hotels or simply left standing.

Seed heads of perennials and stems of grasses can contribute color, texture, and movement to your garden during the dormant season. Once your grasses become windblown and tattered and your perennials have lost their seed heads, you can tidy up, but keep in mind the overwintering insects. In short, cut back on your fall work, not your perennials!

OPPOSITE Switchgrass (*Panicum virgatum*) left standing for the winter provides color, texture, and movement.

RIGHT Leaves raked off the lawn can be used to mulch beds of shrubs and perennials.

Leaf raking

Leaves from deciduous trees are a huge supplier of nutritional benefit. Run over them with a lawn mower or use a leaf shredder to provide more openings for microorganisms to enter and begin the process of decomposing and releasing nutrients. These shredded leaves make a great mulch for garden beds, shrubs, or trees. If you use leaves to smother a lawn over the winter (see page 45, Layers), then shredding is not necessary, as your aim is to keep them intact for as long as possible. Rather than raking, bagging, and working to get rid of every last leaf, I advocate doing less to get more.

Protecting plants from wind and salt

Winter protection is essential for some plants. Cedars or pines planted close to the roadside may need protection from salt sprayed up by passing vehicles; shrubs that are semi-hardy in your zone may need protection from the wind. Unfortunately, it is very difficult to protect plants without that protection becoming the focus of your winter garden. Burlap is the material of choice because it is biodegradable and can be obtained in a loose or tight weave. It allows for some airflow, but blocks a good deal of air and absorbs salt spray before it hits plant foliage. However, it is difficult to attach to supports, has a limited life-span, and tends to stretch out over the winter, looking baggy and bedraggled by spring.

LEFT A mulch of winter greens adds a layer of protection to roots that are vulnerable to heaving when stones heat up and thaw the soil, only to be refrozen each night. RIGHT An enclosed garden of carefully pruned shrubs is protected and also creates a sense of intimacy.

Some newly planted deciduous trees, especially young fruit trees, may benefit from wrapping the bark with a crepe paper–like product to protect them from "sun scald." Young trees with smooth, thin bark are most vulnerable when sun from the south or west reflects off of snow onto tree stems, warms the bark, and stimulates biological activity, only to be followed a few hours later by freezing weather that causes the bark to split. The paper wrap also helps prevent rodents from girdling the bark while the base of tree is under the snowpack.

If you are tired of the work required to protect your plants and unhappy with the way they look over the winter, consider moving them to a sheltered location, where they may not need protection; moving semi-hardy shrubs out of prevailing winds or to a location that warms up next to a building can be helpful. Or try replacing them with species that are better able to cope with the adversities of the site. Replacing semi-hardy plants with hardy natives that fill the same niche can improve your view for the six or seven months of the year when more delicate plants would require protection. By replacing plants that are vulnerable to salt spray with those that can tolerate salt (usually a species native to the seashore), you can reduce your fall chores and better enjoy your winter views.

DEER PROTECTION

Although it may be the antithesis of garden ornament, deer defense is a requirement for those of us who want to protect our gardens from winter browse or just want to keep deer and other animals out of our vegetable gardens year-round. How can we do this thoughtfully and economically?

First, we need to understand the creatures who love our plants as much as we do, but for different reasons. Deer are creatures of habit, often following the same routes for a period of time, browsing the same or nearby plants as they go. They are looking for the most nutritious foods, which change seasonally according to availability. When perennials are dormant or covered in snow, woody plants and shrubs start looking better. The most nutritious part of the plant is going to be the growing tips: buds of deciduous plants and needles of conifers. If more than just the tips are being browsed, that is a sign that the deer are having trouble finding enough good food and that they are willing to eat just about anything! Deer are adaptable. They typically feed in the early morning and afternoon; the rest of the time they are digesting and sleeping.

To distinguish between deer and rabbit or hare browse, check to see if the remaining stub is rough or smooth. Rabbits have upper incisors and can bite off stems at a sharp angle, which looks like a knife or scissor cut.

LEFT A tall fence with an openwork gate keeps out deer but welcomes humans.
RIGHT Sometimes a simple deterrent is enough.

ABOVE LEFT Seven-foot-high plastic deer fencing with a whimsical gate protects a vegetable garden.

ABOVE RIGHT This beautiful openwork fence is removable to allow access for vehicles.

RIGHT Deer and rodent protection is necessary for vulnerable specimens.

Deer, on the other hand, lack upper incisors and leave a frayed, ragged edge. The height of the browse may also be a clue.

Next, we need to decide whether to exclude deer or repel them. One very good option for exclusion is an electric fence. I have found it very effective to surround an area with permanent or temporarily erected posts strung with several courses of wire. (I recommend wire rather than fiberglass tape, which is both more visible and less effective.) Attach a solar fence charger to at least the wire at feeding height (about 4 feet [1.2 meters] off the ground), bait it with peanut butter smeared on aluminum foil, and you are all set. The shock the deer receive is harmless but memorable. The highest wire should be about 7 feet (2.1 meters) off the ground for best protection. It may not look great, but it will do the trick. Another option is to permanently enclose an area with a decorative fence that you can extend vertically in the winter—or you could add an electrified wire at feeding height to the outside of the fence for the winter.

It is also effective to exclude the deer from just the few plants that both you and the deer find attractive. This can be done with wire fencing formed into a cylinder and hung on metal or wooden stakes around each tree or shrub. Heavy-duty black plastic fencing with a 1-inch (2.5-centimeter) grid is even more unobtrusive and lightweight for easy summer storage. It comes 7 to 8 feet (2.1 to 2.4 meters) wide. I use this for cedars, which need protection all the way to the ground, as they are a favorite treat for deer during the winter. Bird netting may be tempting because of its reasonable price and compact storage, but is difficult to use and if you don't remove it soon enough in the spring, plants will begin leafing out through it, making it impossible to remove without injuring the plants. Burlap is another option—especially if you want to protect shrubs from winter winds as well as from deer—but it is harder to set up (you will need to staple it to wooden stakes or wrap it around a wire fence structure) and it has a very limited lifespan and can rarely be used for more than one winter.

Install hardware cloth protection at a height of about 36 inches (91 centimeters) around the trunk of new trees to prevent bucks from rubbing their antlers and disfiguring young trees. A shorter version will also deter voles and mice from girdling the trunk by eating the inner bark while the base of the tree is under the cover of snow. (For voles, a height equal to the depth of snow you anticipate is adequate.) Anchor the base 1 inch (2.5 centimeters) or so below the surface. Tree wrap can also be an effective deterrent.

Repellents can be another good option if you are committed to using them year-round. While I don't recommend spraying liquids during freezing weather, I have found that if you are successful in deterring deer throughout the rest of the year, they will come to think of your garden as off-limits. (Of course, all bets are off in a particularly difficult winter or with an unsustainably large herd of deer.) Most of the products on the market that contain putrescent eggs work for several weeks. I reapply according to label directions or more frequently after spells of hard rain. If deer seem to be holding their noses and eating anyway, add a product containing capsaicin—the ingredient that causes the heat in hot peppers. Smelly soap hung in onion bags from branches or tied to stakes can also deter deer. Mixing up your strategies is always a good idea.

Think about the vulnerability of your plants to deer and have a strategy for protecting each one. When you read that a plant is deer-proof, that might mean something to the person doing the writing but unless it means something to the deer, your plant may be in trouble. The only way to know is to experiment, and, as we all know, what works one year may not work the next.

PRUNING BASICS

Winter is a great time to prune most deciduous shrubs and trees. Pruning can feel scary—there are so many rules. What if you do something "wrong"? The good news is that when you prune in the winter, you are least likely to do any damage; it is also the time of year when you can really see what you are doing.

Your first question should always be: Why should I prune? In many cases, pruning is not necessary, especially if your plants are native to your region and you have placed them appropriately. However, top among the good reasons to prune is plant health. Check your trees and shrubs for dead and damaged branches and prune them out. Make your cuts as close

LEFT Neatly done deer and rodent protection needs not be obtrusive.

BELOW LEFT Bamboo fencing in combination with an electric wire can be an effective deer deterrent.

BELOW RIGHT Carefully pruned and tended shrubs retain their lower limbs when protected from browsing deer.

A late winter snowstorm interrupted the pruning of this crabapple.

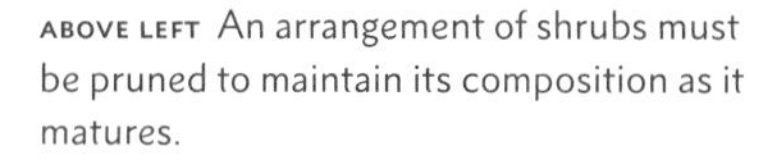

ABOVE LEFT An arrangement of shrubs must be pruned to maintain its composition as it matures.

ABOVE RIGHT A carefully pruned tree close to the house will remain beautiful and in scale for many years.

RIGHT Japanese maples have been bred to be all shapes and sizes. Be sure to investigate the growth habit before investing in one.

to the junction between the damaged branch and a healthy stem as you can while leaving a bit of the swelled area at the junction.

Pruning to maintain the size of a plant

Reading about the ultimate height and girth of a plant is helpful when doing your initial planting, but woody plants are living things and continue to grow throughout their lives—they don't always stop when they reach the "maximum height" on the tag. Remember that you are planting them in artificial conditions and nurturing them as they would not be cared for in nature, so there may be more variability than reference materials would indicate. Maybe your shrubs were spaced well enough for their first ten years, but eventually they may grow into each other or a building or they may shade out their neighbors.

The best strategy to reduce the size of a vigorous shrub is often to remove the tallest stems or the ones that are growing furthest outboard. Prune these to within an inch or two of the ground and let new, smaller stems sprout out. If your plant is not as vigorous, you can remove the top half of the taller stems down to a junction one year. When the other stems have filled in a bit, take the remaining half of those stems out the following year. Pruning in stages is often the best way to achieve a goal that would be too drastic to accomplish all at once. This is also a good way to get to know your plants; observing how each plant responded to your cuts the previous winter can inform your strategy for future pruning. Go slow and avoid pruning out more than one-third of the woody growth at one time.

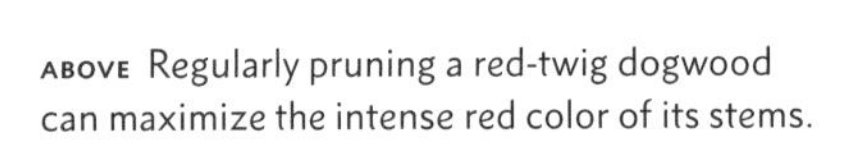

ABOVE Regularly pruning a red-twig dogwood can maximize the intense red color of its stems.

RIGHT Pruning trees and shrubs to complement each other requires an understanding of their natural tendencies.

This Stephanandra (*Stephanandra incisa*) needs regular pruning to keep it from overgrowing the bench.

Pruning dogwoods and willows for winter color

While many resources suggest pruning dogwoods and willows to the ground each year for best color, this is not advisable in our northern climate unless you know the root system is strong and will respond vigorously. You can test this by taking out several older stems from each plant (these will be the least colorful or the most gray) and observing the regrowth. If the regrowth of cut stems is only about 12 inches (30.5 centimeters) and you want a taller shrub, you will have to prune in rotation each year. If regrowth after pruning is 3 feet (0.9 meters) or more, you can cut more stems each winter. Again, go slow, observe as you go, and adjust your practice.

Pruning for lovely winter shape

Just as the structure of the garden is more obvious in the winter, the structure of each tree or shrub is most apparent without leaves, making decisions about which stems or branches to take out a lot less difficult. It is easy to be pruning away in the summer and not notice that the branch or stem

ABOVE These Sargent crabapples have been trained into an arch and pruned to maintain an open structure. **OVERLEAF** The Japanese-inspired Asticou Azalea Garden in Northeast Harbor, Maine, was designed by Charles Kenneth Savage in 1956 and is impeccably maintained, demonstrating the beauty of thoughtfully pruned plants at every turn.

you are leaving has a split or a wound that you didn't see because of the foliage. Most of the pruning you do in the dormant season will encourage new growth at the cut. Make a point of dealing with unwanted suckers by pruning out all or most of the weak shoots that sprout out from latent buds near the cut during the summer, after growth has slowed down and they are not as likely to resprout.

So, what shape are you going for? Unless you are pruning to maintain a formal garden feature like a hedge, use nature as your guide. Look around your area, on roadsides and in natural areas and parks, to observe the structure of shrubs and trees you find attractive. For me, it is often the openness of a shape that appeals. By pruning out a lot of interior twigginess, you can get a simpler, more elegant form. This is pretty much the opposite of the way pruning is usually done, where the goal is to make plants look full. But in the winter, fullness is usually not an option, and a beautiful structure can be appreciated through a plant's foliage in the summer almost as well as it can be admired during winter dormancy.

I would recommend shearing, where all the branches are cut to the same length, only in the case of a formal hedge or a coniferous tree or shrub. Even then, it must be used judiciously, and occasionally pruning out the often sheared and frequently branched tips is usually a good idea to promote air circulation and light penetration for the health of the plants. I call this "thinning," and I do it in the summer as well as the winter to promote a vigorous and healthy shrub or tree.

There are many books devoted to pruning, and I encourage further reading, but it is not strictly necessary as long as you keep in mind the few basics I have mentioned here before heading out. I can't stress enough the benefit of having a quality pair of hand pruners with a sharp blade as well as a small folding handsaw, also with a sharp blade. With these two tools, I can prune just about anything within reach. I once asked an arborist the best time to prune, and he said his grandmother had taught him that the best time to prune is "when your tools are sharp." Clean cuts with no jagged edges are vital to quick healing and getting on with growing when the weather warms. With pruning, each plant will be your teacher if you have the patience to wait for it to speak!

LEFT *Tamamono*-pruned spirea contrast with standing grasses and perennials.

BELOW RIGHT *Karikomi*, or wave-pruned, mugo pines (*Pinus mugo* cultivars) enclose a secret garden.

BOTTOM Boxweed (*Buxus microphylla*) pruned into balls outline the paths in this walled garden.

Multiple magnolias are pruned into a single canopy.

SPECIAL PRUNING TECHNIQUES

Tamamono and karikomi

When you are using shrubs to add topographical relief to the garden, the Japanese pruning style of *tamamono* is useful. Small shrubs are individually pruned in the shape of an inverted bowl. This dome shape, which is wider than it is tall, is suggestive of a boulder protruding from the ground. A grouping of *tamamono*-pruned shrubs where one shrub grows into another is called *karikomi*. Each shrub remains defined, but together the larger form looks like the undulating surface of the sea, so this is also known as a "wave."

While tightly pruned shrubs in these and other forms are evocative of Japanese gardens, these concepts can be used more broadly and with many more shrub species than the typically used evergreen azaleas, yews, and hollies. Pruning a grouping of shrubs into an undulating form or considering an entire group of shrubs as a single form when pruning can add a new element to the garden structure and alleviate the need to see and prune each shrub individually.

Topiary

Topiary can be integral to the structure of a formal garden, a part of the garden architecture, or simply used to shape an individual ornamental plant, akin to a sculpture. From geometric shapes to oversized animals, these designs require a steady hand and a commitment to training your plants. They often look best in formal garden situations, but don't let that stop you from using your creativity to enliven your garden with living sculpture. A dusting of snow can add a new dimension to your creations, but be cautious about creating fragile, overhanging shapes that will be vulnerable to breakage under a snow load.

Espaliered cotoneaster (*Cotoneaster adpressus*) adorns this wall.

This apple tree has been animated by regular pruning.

Espalier

Espalier is an ancient method of training trees into a vertical plane, using a trellis or some other means of support. It can embellish a wall or create a freestanding fence or hedge that is barely more than 12 inches (30.5 centimeters) deep. This is a great technique if you have a small garden or if you would like a fence but prefer something living. Apple or pear trees are frequently used to create fruit-bearing ornament, but many other species with attractive flowers or fruit can work, too. Start with whips; plant them in a row, cut off the top, and begin training the side branches along a trellis or wire to either side. Remove branches growing out or back toward the trellis or wall.

Pleaching is similar to espalier, but with the trunk of the tree bare and the foliage beginning at or above head height. The result is something of a hedge on stilts. Like espalier, the depth of the structure is often narrow, but the impact of the pleached trees is magnificent. Consider this technique if you need a high screen or divider.

Editing

Another kind of pruning is analogous to editing. I use this for transitional areas that could use aesthetic improvement but don't want to be gardened. A little editing of branches, foliage, and stems can present a neat appearance without looking artificial or fussy. The edge of a driveway, woods, or meadow can be enhanced by editing out plants or foliage that are crowding more desirable plants.

For example, I have some lovely rhodora (*Rhododendron canadense*) plants that I put in between my driveway and the woods. I don't want to turn this area into a "garden," but I do want it to look neat and to appreciate the rhodora during the winter, when the flower buds are prominent, and especially when it is blooming at the end of May. To highlight the rhodora, I try to keep the other plants in the area under control. I have repeatedly cut back alders, seedling maples, and even the rushes that sprout up in this wet area. Note that I do not remove the plants, just cut them back to the ground or an inch or two above. This has put a dent in the enthusiasm of the alders, now that I have been treating them this way for a couple of years. (Don't worry, I don't cut back other alders that provide early spring pollen for the bees in a different area.) Before the rhodora were large enough to fill the space, I left some of the resprouting alders to provide foliage and structure between the rhodoras, preventing them from looking lonely and "planted" while contributing to the shrub fringe on the edge of the woods.

Make sure the type of pruning you do reflects the degree of formality of the area surrounding the pruned subject. While editing can be used most anywhere in a garden to maintain a natural aesthetic, the more highly stylized methods discussed earlier should be reserved for plants in a formal situation. As always, consider the balance between nature and human-made and where you want your garden to lie on that spectrum.

Manicuring a wild edge by removing, pruning, and shaping can ease the transition between cultivated and uncultivated spaces.

FIVE

PART FIVE

SHARE

Making the Most of Your Garden during the Quiet Season

The Scandinavians are experts at taking winter in stride, as demonstrated by the Danish term *hygge* and the Norwegian *friluftsliv*, both of which have lately garnered worldwide usage. Neither of them, however, have a direct English translation—maybe because these concepts are not typically a part of the American way of life. *Friluftsliv* is roughly translated as "outdoor or open-air living." It embodies the custom of making a conscious effort to get outside on a regular basis and to connect with nature. It is only recently that the actual physical benefits of being outside have begun to be quantified—something that has been known by the Scandinavians for hundreds of years. *Hygge*, in contrast, is all about coziness and intimacy—creating rituals that instill familiarity and contentedness.

Both of these concepts can apply to the winter garden. While outdoor living and coziness may seem to be at odds with each other during the winter, think again. Don't we appreciate candlelight more in the darkness or the warmth of a fire when it is cold outside? The contrast of the expectation of chill and darkness with the warmth of a hand-knit scarf and laughter with friends around an outdoor fire makes sense of both terms. Here's to sharing your garden with friends and family, as well as with the other creatures who make your garden their home all year long.

TOP LEFT Flowerheads drying in a toolshed.

BOTTOM LEFT Seed capsules of red vein (*Enkianthus campanulatus*) add color and texture all winter.

BELOW A simple decoration indicates that this is the home of a gardener.

BRINGING THE OUTDOORS IN

If you are like me, you begin scouring your garden in early December for attractive greens, berries, and seed and flower heads to bring inside for holiday decorating. It's not surprising that the plants you bring inside to appreciate during the winter months are the same ones that also look great in the garden. So why not plant your own supply of decorative winter plants? Here are some plants that have specific characteristics that make them desirable for indoor or outdoor decoration.

Berries

Some late-ripening berries, drupes, or fruits retain their color and shape at least through the early part of the winter (or until a pack of hungry cedar waxwings discovers them). Winterberry holly (*Ilex verticillata*) and the many varieties of evergreen holly (*Ilex × meserveae, I. opaca*) have bright red berries that are occasionally overlooked by the birds. Our native roses such as *Rosa carolina* or *R. virginiana* have hips that remain an attractive mahogany color borne on red stems. Other highly decorative shrubs or small trees include chokeberry (*Aronia arbutifolia*), highbush cranberry (*Viburnum opulus var. americanum*), mountain ash (*Sorbus* spp.), crabapple (*Malus* spp.), and Korean dogwood (*Cornus kousa*).

Seed and flower heads

Even if you have a stock of mid- and late summer blooming perennials and grasses that you leave standing throughout the winter, it sometimes makes sense to cut them earlier if you intend to use them indoors. You might consider cutting poppies and many grasses and sedges early, standing them upright or hanging them upside-down in a well-ventilated area like a potting shed or garage to dry for use months later. While some perennial flower heads such as echinacea and teasel look great when they dry to a chestnut brown, the seed heads of many poppies and alliums will dry to more of a straw color if cut early. Not only will they contrast with darker shades of evergreen foliage and flower heads, but you can eliminate reseeding of prolific self-seeders you already may have too many of. Some trees have fantastic seeds, including silver bell (*Halesia carolina*) and hornbeam

(*Carpinus* spp.), and some shrubs such as buttonbush (*Cephalanthus occidentalis*) and summersweet (*Clethra alnifolia*) do, too. Flowers from the many hydrangea species and cultivars look great if cut early and dried. Why not plan to cut some early and leave the rest?

Vines

Vines are useful for wreath making and as the structure for garlands, swags, and other holiday decorations. If you intend to use them for wreaths, it is best to shape them immediately after cutting, while they remain flexible. Attractive and sturdy vines typically come from perennials including grape, wisteria, Virginia creeper (*Parthenocissus quinquefolia*), Dutchman's pipe (*Aristolochia durior*), and trumpet vine (*Campsis radicans*). Vines with other attractive features include wild cucumber (*Echinocystis lobata*), which has amazing seed pods, and virgin's bower (*Clematis virginiana*) and other clematis, which have feathery seed heads.

If you find yourself removing invasive Oriental bittersweet vines in the summer, strip the foliage and wind them into a wreath to save for winter decoration. There are probably plenty of land trusts and landowners who

Repetition makes an impact.

FAR LEFT A small sprig of red-twig dogwood (*Cornus sericea*) can be used indoors as well as out to welcome visitors.

LEFT Harry Lauder's walking stick (*Corylus avellana* 'Contorta') is a clone that produces branches suitable for year-round flower arranging and decorations.

ABOVE A grapevine wreath adorned with winterberry.

would welcome folks to harvest these nuisance vines before they take down their trees. Do not, however, be tempted to use the attractive berries—they will drop off on your way to the house or be picked off your wreath by hungry birds and continue their invasion! A few other invasives to avoid are *Akebia* and Japanese honeysuckle.

Twigs

The shrub dogwoods are famous for their brightly colored twigs. Willows also come in shades of red, yellow, and chartreuse. The key to obtaining the brightest twigs is to use the recently sprouted stems from the prior year's pruning. Even red maple will send up stump sprouts in a deep red that can be cut and used for winter appreciation. Twigs with interesting textures can also be used like those from larch (*Larix laricina*) or those with fascinating

FAR LEFT Black pussy willow (*Salix melanostachys*) can be striking.

LEFT The husks of beech nuts remain on the tree long into the winter.

ABOVE Cones of evergreens are unique to each species or variety on which they grow.

shapes such as Harry Lauder's walking stick (*Corylus avellana* 'Contorta'). Plants with colored foliage like purple-leaved smoke bush (*Cotinus coggygria* 'Royal Purple') can be lightly pruned in summer and dried to retain their colorful foliage. The stark white bloom on the stems of blackberry twigs can also be striking both in the winter garden and in arrangement, but handle with care!

Foliage

Of course, boughs from any evergreen will be welcome at this time of year. Spruce, fir, pine, cedar, hemlock, or yew can contribute color and texture to your winter decor. Of the broad-leaved species, these plants with their attractive deep-green foliage retain their beauty for a time after cutting: hollies (*Ilex glabra*, *opaca*, and × *meserveae* hybrids), dog laurel (*Leucothoe*), andromeda (*Pieris* spp.), and boxwood (*Buxus* spp.). Try using the trailing, fine-leaved foliage of bearberry (*Arctostaphylos* spp.) or one of the cranberries (*Vaccinium macrocarpon*). Rhododendron hybrids and mountain laurels (*Kalmia latifolia*) maintain their leaves throughout the winter but tend to droop when temperatures drop as a mechanism to prevent excessive transpiration. They do remain green but look rather sad in the deep cold.

Catkins

Some of our native trees and shrubs retain their female catkins throughout the winter. Members of the birch family, which includes hornbeam, hazelnut, and alder, are festooned with nut-brown catkins throughout the winter and may even be sporting their colorful, unopened male catkins in preparation for spring. In fact, they will be dispersing seed throughout the winter that will germinate in the spring. (As an aside, members of this family are some of the earliest woody plants to bloom in the spring. This early nectar and pollen are valuable resources for our native pollinators and honeybees—yet one more reason to plant these in your garden.)

Cones

Spruce, pine, fir, and hemlock cones are always useful around the holidays. Collect them from the ground after they have dried, opened up, and released their seeds. If you do bring mature, unopened cones into a dry place, the complex design of their cone scales will force them to open. Cones of some species remain on the trees after opening; these can be clipped off—just leave some on the trees. Hungry birds and squirrels will find the current year's seeds nestled within!

Acorns and such

There are as many as ninety different oak species in North America, making them one of our most abundant and diverse trees. They all produce acorns, which pack a lot of food for newly germinated oak trees, making them both attractive and nutritious to wildlife. Other nuts (some are technically fruits) that fit this bill are beech nuts, hickory nuts, chestnuts, and horse chestnuts. Nut producing trees usually don't begin bearing until they are twenty years or older and will eventually become tall canopy trees, so plant with this scale in mind.

Think creatively about combining various plant materials to make a stunning decoration.

Birch bark

You can remove the loose outer bark from mature paper birch trees at any time without damaging or defacing the trees. This requires no tools and yields a thin, flexible, attractive, paperlike sheet. I do not advise peeling away a thickness of bark that requires a knife to loosen: removing bark that you intend to use for structural items like canoes and boxes can kill the tree or leave an unsightly scar.

The seeds of many trees and shrubs are simply stunning (clematis, witch hazel, azalea, maple, and wild yam).

Mosses

Foam balls or other shapes can be covered in moss to use as decorations. Wire the mosses onto the foam core and hang them, attach them to wreathes, or cluster them in a bowl. You can also use twine or thread to attach the moss. Mist your moss balls occasionally to keep them looking fresh. Other natural materials, such as leaves that you collect before they become brittle, can be used in the same way. Soak the dried leaves in water to make them pliable enough to bend into shape.

You can clip dead stems, foliage, or other plant parts at any time after the flowers are finished and the seed pods still look fresh. Experiment with timing, as colors may fade if they are cut too early. Pruning, on the other hand, should happen either during the growing season or after the leaves have fallen and the plants have gone dormant. Pruning too early in the fall can promote growth just when plants should be hardening off for winter, so it is wise to wait as long as possible.

GETTING OUTSIDE IN WINTER

There is a cozy satisfaction that comes from viewing the garden from the inside during the winter, especially during inclement weather. However, consider what you might do to entice your family outside when conditions aren't so dreadful. Here are some ideas for creating a winter feature to draw everyone outside, even on the coldest day.

Firepit

A fire at night can attract people like moths to a light even in frigid weather. A permanent firepit can be as simple as a ring of local stone or as complicated as an actual fireplace. Enhance the feeling of remoteness by locating the firepit at a different elevation from the house or by screening it with a stone wall or plantings such as evergreens or clustered deciduous shrubs or trees. Screening devices will also help to break the wind.

Many types of portable firepits are available. For safety reasons, you should create a stone hearth for the pit by setting it on a small patio designed for this purpose or by placing it on an existing hardscaped area. Sitting around the fire on a cold evening with friends is a great way to celebrate the darkness and the light—a perfect way to spend the winter solstice!

Sauna

Nordic cultures are famous for their outdoor saunas, which are valued for relaxation and socializing. After sitting in the wood-fired heat and ladling water onto a bed of hot rocks for a burst of steam, sauna goers cool off by jumping into a lake, rolling in the snow, or, these days, taking a cool shower. Then repeat! A sauna can be a magnet for like-minded friends who enjoy its cleansing properties.

Locate the sauna near a pond or in a place with some privacy that you won't mind hauling water to. You will also need a sheltered place to stack wood. While saunas can be rather extravagant, kits and plans are available for the adventurous do-it-yourselfer. (Check local zoning ordinances before you begin construction.)

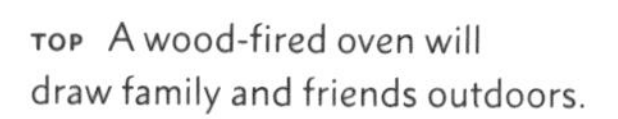

TOP A wood-fired oven will draw family and friends outdoors.

ABOVE The beauty and tradition of sauna are especially welcome in winter.

RIGHT Enjoying fire can become a wintertime ritual.

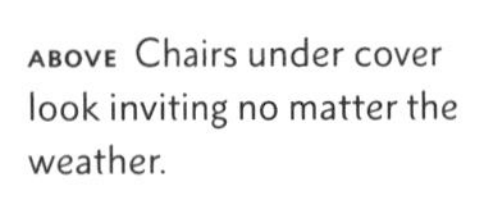

ABOVE Chairs under cover look inviting no matter the weather.

RIGHT An outdoor shower is invigorating.

Hot tub

An outdoor hot tub can be a real luxury in the winter. Imagine soaking up the heat of the water while you search for winter star constellations or watch the snow fall. You may need to do some work to winterize your tub; keep this in mind when shopping among the variety of tubs, kits, and plans. A simple cedar tub with either an underwater or external wood stove to heat the water will alleviate the need for electricity and can be filled using an ordinary garden hose. Like the sauna, your hot tub will entice friends and family outside, even in frosty weather.

A place to tie or untie boots near the door.

Viewpoint

Add a bench at a location that overlooks a special view during the winter months—maybe one that is obscured by leaves during the growing season. Or simply add a log to perch on near your woodpile to watch the squirrels or birds at a feeder. A protected location that catches the warming rays of the sun can be heavenly when it is too windy or cold to sit still elsewhere. Try out a couple of spots using a portable chair before committing to a bench. Once you decide on the location, install a bench or chair that reflects the style of your garden in a weather-resistant material like stone, wood, or a composite. Think about the color of the seating in its winter surroundings. A brightly colored bench might stand out too strongly in the winter—or it might make you smile each time you approach.

RIGHT This whimsical playhouse may be less useful but no less ornamental in winter.

BELOW A permanent labyrinth may not be visible under snow; why not have fun and create a snow labyrinth?

Labyrinth

If you are lucky enough to get a good, solid snowfall, how about constructing a snow labyrinth? All you need is a large area of untrampled snow and a pattern to follow. Walk out into the snow in a winding line that does not cross another line—usually in a circle, starting at or near the outside perimeter and winding your way inside. A spiral works well, too. Walk out of the labyrinth along a parallel line. Encourage others to follow in your footsteps as a meditation, releasing while walking toward the center, receiving at the center, and returning on the way out. Good for stress reduction, reflection, or celebration, these snow creations have the benefit of impermanence. You can make a new and different one each time it snows. If you don't have snow, you can still create a labyrinth by outlining the path (the places you don't walk) in boughs, brush, unmown grass, or—more permanently—in stone. Embellish your creation with candles or something special in the center.

A devoted space is not required to enjoy the outdoors during the winter. I encourage you to spend more time outside, using your senses to experience your surroundings. We talk about the garden mainly in visual terms, but don't forget to feel the warmth of the sun, listen to ice tinkle, and smell the dryness of the cold. While you are out there, find some time to play.

GARDENING FOR THE BIRDS

Birds in winter, like the garden itself, are a more subtle shade of their summer selves. This does not detract from the joy I find in watching them hunt around for seeds or flit from branch to branch or in hearing them vocalize on a clear, sunny day. I know many people share this joy, and you can, too.

The birds that are hardy enough to stick around in northern climates do so because they have adapted to our harsh weather and have figured out ways to find food and shelter. You can make their jobs easier by gardening with plants whose seeds and fruit remain on the stem late into the winter. Some great winter foods are mountain ash (*Sorbus* spp.), bayberry (*Myrica pensylvanica*), winterberry holly (*Ilex verticillata*), choke berry (*Aronia arbutifolia*), crabapple (*Malus* cultivars), viburnums like

TOP Bohemian waxwings discover crabapples that nearly made it through the winter untouched.

ABOVE It's always unpredictable where birds will raise their family.

RIGHT A rustic feeding platform made with paper birch limbs and bark.

Birdbaths can be fitted with electric deicers or refilled daily with warm water.

highbush cranberry (*V. opulus* var. *americanum*), nannyberry (*V. lentago*), and maple-leaved viburnum (*V. acerifolium*), and, if you have the space, staghorn sumac (*Rhus typhina*). The fruits left on these plants through the winter will make them attractive to you, too.

Also, if you refrain from cutting back your perennials and ornamental grasses, their seed will be released to neighborhood birds over the winter. You should see a diverse array of seeds on top of the snow or collecting in footprints or animal tracks. Many birds will harvest these seeds straight off the plants, but some, like juncos and sparrows, prefer to feed off of the ground.

High-protein foods like insects are especially important for birds. By welcoming insects into your garden and avoiding insecticides, you will also be welcoming birds. The larvae of many insects overwinter in the soil, in leaf litter, or in crevices on the bark of shrubs and trees, where they can be found by hungry birds during the leafless months.

Supplemental foods such as birdseed and suet will also be welcomed by our feathered friends when other forage is limited. A variety of foods will attract a variety of birds. Black oil sunflower seeds are considered one of the best foods, being nutritious, high in oil content, and having a thin shell that smaller birds can break open. Often seeds from a standard bird-food mix are wasted by birds trying to get to the sunflower seeds. If you want to attract specific kinds of birds, find out their seed preferences and offer that in an appropriately designed feeder.

Try to be consistent about filling your feeders, keeping them clean—wash at least twice a year with a 10 percent bleach solution—and protecting them from unwanted diners, such as squirrels and raccoons. Install a baffle on the post or hang your feeder over a high tree branch and cleat it off so it can be lowered to refill. Taking your feeders in at night also reduces the possibility of invaders.

Locate your feeder near a shrub or tree to provide the birds with a staging area. A conifer or thickly branched shrub about 10 feet (3 meters) from the feeder will provide needed cover for birds without becoming a launching pad for squirrels.

Suet is a high-energy food that attracts chickadees, nuthatches, and woodpeckers. Use a metal cage if intruders are in the neighborhood; otherwise, a mesh bag is adequate. Use commercially available cakes or make your own. Suet is a cold-weather food—it can get rancid in warmer weather, so take it indoors when temperatures rise above 40° F (4° C).

Water is perhaps the most difficult habitat element to provide for wildlife during the winter if you don't already have a pond, pool, stream, or brook. Think about the larger landscape surrounding your property. Is there a nearby source of water? Only if none exists should you consider using a birdbath heater to keep water thawed during freezing temperatures. Most wildlife are resilient and resourceful and don't need electricity to withstand the winter. But if it makes you feel better to keep some open water in your yard, by all means go ahead. You can purchase a birdbath with a built-in heating element or add a birdbath deicer to one you already have. The advantage of the built-in models is that the cord is concealed in the unit. A birdbath can either be placed on a pedestal or set directly on

LEFT A humorous birdhouse that might be more for looks than for function.

BELOW Keeping water from freezing in a vessel such as this can be a challenge.

the ground; if you have outdoor cats in your neighborhood, the pedestal type will be best. For the committed bird lover, removing the ice from an unheated basin and refilling it with fresh water each day will give the birds at least an hour or two of open water.

One winter, during a blizzard I saw a flicker clinging to the top of a bayberry (*Morella caroliniensis*) to dine on its wax-covered seeds. I felt that just by planting that bayberry I had made a miserable day a little better for the flicker. The sight filled me with compassion for the bird and reinforced my mission to create a garden that everyone can enjoy.

TOP LEFT Barred owls are sometimes spotted during the day but may be using your garden more than you know at night!

TOP RIGHT In Maine, we enjoy seeing our spunky red squirrels throughout the winter.

BOTTOM LEFT A squirrel has commandeered this nest box and turned it into a larder.

BOTTOM RIGHT A recently constructed insect hotel provides nesting materials and a home for bugs and birds.

OPPOSITE Leaving standing vegetation provides nesting areas and material for garden creatures throughout the winter.

CREATING HABITAT

Gardening for wildlife sounds like something we all want to do, but when people talk of insects in the garden it is mainly about how to eliminate them. However, each year new research reveals how beneficial insects protect your plants from more problematic species such as aphids, cutworms, even slugs and, yes, Japanese beetles, to name a few. And concern for our native pollinators is peaking now that we are seeing fewer and fewer bees, butterflies, and moths in our gardens and natural areas.

All insects need places to overwinter if you want to enjoy their benefits in the summer. Safe places for eggs, larvae or pupae, and adults awaiting spring are necessary. While each species has its own particular requirements, here are some simple things you can do for these garden helpers:

- Avoid tilling in the fall and early spring; many find homes in the soil.
- Avoid raking up every last bit of leaf debris from your garden in the fall; many need leaf mulch for winter protection.
- Leave as much standing vegetation as you can bear; many lay eggs in hollow stems.
- Cherish your mature trees and consider leaving them standing even if they are dead; they provide habitat for thousands of species of insects, as well as lichens, mosses and fungi.

Another strategy is to build some brush piles that can provide habitat and additional shelter. Loose bark on dead limbs can become a home for invertebrates or even tree frogs. Spaces within the pile provide cover, and the composting material beneath is ideal for frogs, toads, and small mammals to burrow and nest in. These piles should be in place by the time temperatures begin to drop in the fall but before the ground freezes. Wouldn't you rather small mammals find homes in your garden than in your toolshed or garage? Check around your piles after a snowstorm to see who is coming and going; most field guides to mammals will show you the sizes and patterns of footprints made by the creatures you are most likely to see. Amphibians using your pile will stay put for the duration. With a little creativity, a brush pile can actually be a feature in your garden—a structure for snow to fall on, something to cast shadows in the low winter light, or an amusing sculpture to put a smile on your face. As the pile breaks down, it will become home to fungi and other organisms that are mostly beneficial to your garden plants. You can either add to the same pile each year or start a new one. After a year or two, the soil beneath each pile will be energized with microorganisms and ready to feed a new garden bed or tree.

Piles of stone or dry-stacked stone walls also provide shelter for insects, amphibians, and reptiles. Again, be creative—a pile of stones can be artfully done! If you are lacking appropriate space or think these piles may look too messy, you can build a "bug hotel" in a protected wooden box. Include materials with nooks and crannies for winter protection, such as bricks, logs, brush, hollow twigs, and soft nesting materials like wool or drier lint. Some creatures will overwinter in the hotel; others will take nesting materials to a place of their own choosing.

Planting native herbs, shrubs, and trees in your landscape is probably the best way to assure your garden will have the habitat that creatures need. These plants will provide shelter, nesting material, and nest sites whether standing or dead. All our native trees, shrubs, and herbs produce foliage, pollen, nectar, seeds, nuts, or berries to feed creatures large and small. And, of course, by avoiding pesticides and all kinds of garden chemicals (most are indiscriminate in their killing), you will be making your garden a better habitat for everyone, including you and your family.

TOP Our native Carolina rose is not only colorful but provides food for birds, nesting material for native bees, and cover for small mammals. No one yet knows how important this one plant is for other species.

BOTTOM LEFT Assembled with creativity and a sense of humor, this brush pile is functional and looks great.

BOTTOM RIGHT Nooks and crannies in this stone wall provide habitat for amphibians as well as overwintering insects.

OVERLEAF Radiating light and warmth, a fire is perfect for an outdoor winter gathering.

Resources

Books

—

Bitner, Richard L. *Designing with Conifers: The Best Choices for Year-Round Interest in Your Garden*. Portland, OR: Timber Press, 2011.

Bloom, Adrian. *Gardening with Conifers*. 2nd ed. Buffalo, NY: Firefly Books, 2017.

Brickell, Christopher, and David Joyce. *Pruning and Training: A Fully Illustrated Plant-by-Plant Manual*. US ed. London: Dorling Kindersley, 1996.

Cullina, William. *Native Trees, Shrubs, and Vines: A Guide to Using, Growing, and Propagating North American Woody Plants*. Boston: Houghton Mifflin, 2002.

Darke, Rick, and Douglas W. Tallamy. *The Living Landscape: Designing for Beauty and Biodiversity in the Home Garden*. Portland: Timber Press, 2014.

Dirr, Michael A. *Dirr's Hardy Trees and Shrubs: An Illustrated Encyclopedia*. Portland: Timber Press, 1997.

Goldsworthy, Andy. *Andy Goldsworthy: A Collaboration with Nature*. New York: Abrams, 1990.

———. *Wall*. New York: Abrams, 2000.

Hendy, Jenny. *A Gardener's Guide to Topiary: The Art of Clipping, Training and Shaping Plants*. Cambridgeshire, UK: Annes, 2018.

Kellert, Stephen R. *Nature by Design: The Practice of Biophilic Design*. New Haven, CT: Yale University Press, 2018.

Lee-Mäder, Eric, Jennifer Hopwood, Lora Morandin, Mace Vaughan, and Scott Hoffman Black. *Farming with Native Beneficial Insects: Ecological Pest Control Solutions*. North Adams, MA: Storey, 2014.

Schenk, George H. *Moss Gardening: Including Lichens, Liverworts, and other Miniatures*. Portland: Timber Press, 1997.

Tallamy, Douglas W. *Bringing Nature Home: How You Can Sustain Wildlife with Native Plants*. Portland: Timber Press, 2009.

Turnbull, Cass. *Cass Turnbull's Guide to Pruning: What, When, Where, and How to Prune for a More Beautiful Garden*. Seattle: Sasquatch, 2012.

Xerces Society. *Attracting Native Pollinators: Protecting North America's Bees and Butterflies*. North Adams, MA: Storey, 2011.

Websites

—

Dirt Simple blog
https://deborahsilver.com/blog/
Inspirational ideas for outdoor displays; check out the posts for winter months.

International Dark-Sky Association
https://www.darksky.org/about/
Recommendations for using night lighting respectfully.

Missouri Botanical Garden Plant Finder
http://www.missouribotanicalgarden.org/plantfinder/plantfindersearch.aspx
Great information on habit and growth form.

Wirtz International Landscape Architects
https://wirtznv.com/projects/
Designers who have revolutionized the use of hedges around the world.

Xerces Society for Invertebrate Conservation
https://xerces.org/publications
A very good source for understanding the value of insects in the garden and creating habitats for them.

Espalier

—

Fletcher, Patricia. "How to Espalier Apple Trees." *Mother Earth News*. October/November 1993. https://www.motherearthnews.com/organic-gardening/espalier-apple-trees-zmaz93onztak.

Thevenot, Peter. "How-to Espalier." *Fine Gardening*, https://www.finegardening.com/article/espalier.

———. "River Roads Farm: An Espaliered Tree Nursery Tour." YouTube video. https://youtu.be/MjGNgLO9_Tk.

Ideas for fun in the snow

—

"Snowlandia in Zakopane." Discover Zakopane https://discoverzakopane.com/snowlandia.html.

"Snowlandia—World's Largest Snow Labyrinth in Poland." The Mind Circle. https://themindcircle.com/snow-labyrinth/.

"Simon Beck's Snow Art." Facebook page. https://www.facebook.com/snowart8848/.

"Snow Drawings at Rabbit Ears Pass, Colorado, 2012." Sonja Hinrichsen. http://www.sonja-hinrichsen.com/portfolio-post/snow-drawings-at-rabbit-ears-pass-colorado-2012/#1.

Acknowledgments

We are grateful to all the gardeners, homeowners, and designers who inspired us with their imagination and creativity and, without exception and on very short notice, welcomed us into their personal spaces. Often our request for a winter visit was answered with "There isn't much to see right now," but, in fact, there was; good design shines through even on the bleakest winter day.

Special thanks to Leslie, for creating a garden sanctuary for birds and humans, where all who enter find sustenance; to Bob, for the guided tour in a blizzard, and to Bob and Jill, for cultivating a garden that continues to inhabit our thoughts, awake and in sleep; to the Wolfs, for answering the knock on the door, allowing us into a favorite garden that lives like it has been loved for generations; to Jan, for believing we would make a good team; and to our friends, family, and colleagues for their suggestions, connections, and encouragement.

Our appreciation extends to the staff and managers of the gardens shown in this book that are open to the public, for both their welcome and their commitment to preserving truly beautiful gardens for generations to enjoy:

- Bedrock Gardens: 9 High Road (location); 45 High Road (mailing address), Lee, NH 03861-6202 | (603) 659-2993 | www.bedrockgardens.org
- Camden Public Library and Amphitheatre: 55 Main Street, Camden, ME 04843 (207) 236-3440 | www.librarycamden.org
- Garland Farm, owned and run by the Beatrix Farrand Society: PO Box 111, Mount Desert, ME 04660 | (207) 288-0237 | www.beatrixfarrandsociety.org
- John Hay Estate at The Fells: 456 Route 103A, PO Box 276, Newbury, NH 0325 | (603) 763-4789 x3 | info@thefells.org
- The Land and Garden Preserve (owner and steward of Asticou Azalea Garden and Thuya Garden): PO Box 208, Seal Harbor, ME 04675 | (207) 276-3727 | info@gardenpreserve.org; Asticou Azalea Garden: 3 Sound Drive, Northeast Harbor, ME 04662; Thuya Garden: 15 Thuya Drive, Northeast Harbor, ME 04662
- Merryspring Nature Center: 30 Conway Road, PO Box 893, Camden, ME 04843 (207) 236-2239 | www.merryspring.org

This book would not be possible without the generosity of the many garden owners who permitted their private spaces to be photographed. Thank you for sharing!

Jeffery and Hillary Becton; Tammy Bernard; Blue Hill Country Club; Bonnie Bochan and Guillermo Diaz; Ann and Paul Breeden; Carol and Bob Calder; Camden Maine Stay Inn; Bonnie Chase; Jennifer Chase and Nigel Chase, in memory of their mother, Susan; Lynn Cheney; Leslie Clapp; Ginger and Richard Dietrich; Linda Elder; Vicki Goldstein; Jan Hartman; Hedgefield, Northeast Harbor; Barbara and Richard Leighton; Heather McCargo, Wild Seed Project; Jan and Tom McIntyre; Jill Nooney and Bob Munger; Ogier Hill Farm; Donna Parratt and Hedgerow Design; Joan Schlosstein; Joanne and Carsten Steenberg; Tom and Yolanda Stein; Peter and Marcia Stremlau; Emily and Robert Stribling; The Bay School; Drs. Thomas Wolf and Dennie Palmer Wolf; Debra and John Piot, who acknowledge their garden design collaborators: Claire Ackroyd, Robin Kruger, and gardeners Guillermo Diaz (head gardener), Juni Charlton, Josephine Jacob, Sarah Schneider, and Mary Cevasco

Published by
Princeton Architectural Press
202 Warren Street
Hudson, New York 12534
www.papress.com

Printed and bound in China
24 23 22 21 4 3 2 1 First edition

Editors: Jan Hartman and Sara Stemen
Designer: Paula Baver

Library of Congress Cataloging-in-Publication Data
Names: Rees, Cathy (Horticulturist), author.
Title: Winterland : create a beautiful garden for every season / Cathy Rees.
Description: Hudson, New York : Princeton Architectural Press, 2021. |
Includes bibliographical references. | Summary: "A practical,
accessible, and lushly photographed guide to making your garden a place
of beauty and inspiration during the winter months as well as throughout
the year" —Provided by publisher.
Identifiers: LCCN 2020055366 | ISBN 9781616898724 (hardcover)
Subjects: LCSH: Winter gardening. | Winter gardening—Maine.
Classification: LCC SB439.5.R447 2021 | DDC 635.09741—dc23
LC record available at https://lccn.loc.gov/2020055366